Management Statistics

Management Studies Series

under the editorship of E. F. L. Brech, B.A., B.Sc.(Econ.), F.B.I.M.

Integrated Managerial Controls
R. O. Boyce

Management Diagnosis—A Practical Guide
R. O. Boyce and H. Eisen

Organisation—The Framework of Management (Second Edition)
E. F. L. Brech

Construction Management in Principle and Practice
E. F. L. Brech

Appraising Capital Works
E. J. Broster

Management Statistics
E. J. Broster

Planning Profit Strategies
E. J. Broster

Co-ownership, Co-operation and Control
P. Derrick and J.-F. Phipps

Marketing for Expansion and Europe
C. Godley and D. Cracknell

Personnel Administration and Industrial Relations
J. V. Grant and G. J. Smith

Marketing for Profit (Second Edition)
L. Hardy

Management Glossary
H. Johannsen and A. B. Robertson

The Economics and Management of System Construction
G. Leon

Management in the Printing Industry
C. Spector

Management in the Textile Industry
The Textile Institute

Industrial Marketing Management and Controls
L. A. Williams

Management statistics

E. J. Broster

Longman

LONGMAN GROUP LIMITED
London

Associated companies, branches and representatives throughout the world

First published 1972

ISBN 0 582 44583 3

Set in 10/12 pt u/lc Plantin 110
and printed in Great Britain by
Adlard & Son Limited
Bartholomew Press, Dorking, Surrey

Contents

Preface

Statistics play a central role in the functions of management and control. In large undertakings, they have done so for many years; but now the new management has brought the truth home to the medium and smaller firms: one cannot manage without statistics. In consequence managers at all levels have to be as numerate as they are literate.

This work is designed to provide a practical guide to those statistical techniques that have been found to be of use in business management, and to their application to the kinds of business problems in the solution of which statistics provide the principal means. On the whole, the early chapters expound techniques, and the later chapters demonstrate some of the more difficult problems to which the management statistician is likely to be called upon to find the answer, and which both managers and management statisticians need to have an understanding of.

One of the techniques expounded is the statistical schedule, which provides a simple means of presenting to top management the significance of an algebraic equation or two or more equations used in tandem, such as the annual cost and annual revenue equations used together for determining profit. The statistical schedule has been known to economists for a century or more. They have used it instead of or as well as a graph, for showing the quantity in demand for each of a range of prices of a commodity. Statisticians have also used the device for comparing the estimates derived from an equation with the actual data on which the equation has been calculated by regression analysis, with the object of determining errors of estimate. But the value of the systematic application of the device by statisticians for presentation purposes has until recently gone unrecognised. It is especially valuable where an optimum is involved in a range of choices for management decision as for instance in price fixing, ordering quantity, and production rates.

Another form of the statistical schedule is the decision tree so called. Here, the diagrammatic tree with its main branches, each with its complex of secondary branches, can be regarded as the academic picture on the box. Presumably the diagram is intended to make the concept easier to understand; but it seems to me it merely serves to lend the concept a mysticism which practical men and women find irritating and wearisome. Inside the box will be found a tabulated form of schedule which management statisticians would have no difficulty in understanding.

Many good management ideas emanate from the universities and schools of business studies, the decision schedule underlying the decision tree being one of them. But many of them, like the decision schedule, are so thoroughly wrapped up in academic mysticism that the good in them is often entirely lost to view. Academic picture and jargon stripping has become a management technique in its own right. It is one that management statisticians in particular will find well worth practising.

As implied above, this work is severely practical. That is not to say it is for beginners. A glance at the table of contents will show that it covers a wide field from the elementary to the more advanced methods. Being severely practical means, amongst other things, that anyone seeking a treatise on the theory of statistics should look elsewhere. Not that theory can be entirely avoided. Statistical theory is concerned with probable error. With the growing recognition that risk and uncertainty are inherent in business decisions, calculating errors of estimate and determining the probable result, together with the most pessimistic and optimistic forecasts from a given course of action, form a necessary function of the management statistician. There was a time when the business manager who had just been presented with a forecast and its probable error, would ask, 'Now you have given me the margin of error, what do I do with it?' It is a question that is not, even now, always easy to answer. The management statistician can often avoid having such questions asked at all by presenting his forecasts, not in the standard form of the most probable figure plus or minus the estimated error for a given probability, but in the form of three figures, viz.: (*a*) the most pessimistic, (*b*) the most probable, and (*c*) the most optimistic. Whether he should state the percentage probability of the two stated extremes depends upon the recipient's knowledge and understanding of statistics.

It has been argued that risk and uncertainty are two different things:

that risk has an objective measure, and uncertainty only a subjective one. There is no reason at all why anyone should not choose to define the two words in that way. That being so, then I submit that in business, risk is rare and uncertainty common—that, in effect, an objective measure rarely exists, though, to be sure, it is easy enough to dress a subjective judgment in objective clothing. It is a danger that constantly faces the management statistician. No matter how sophisticated the tools he chooses to use, no matter how complex the mathematics he applies, neither can provide a satisfactory substitute for basic information. Of course, where a forecasting equation is derived from historic data, there is good reason for calculating the standard error, but it should not be presented as being applicable to forecasts made on the basis of the equation. Forecasting equations are necessarily stochastic in character: they do not take account of all factors. Indeed, it is because they do not, that error exists. The existence of unknown and unmeasurable factors, and therefore of error, provides the be-all and end-all of the science of statistics as distinct from pure arithmetic. If all factors were known and measurable, we would be able to solve most of our numerical problems by employing nothing more sophisticated than schoolboy algebra.

Whatever else a good management statistician may be, he has a number of attributes and qualities, which may be briefly stated.

First, he endeavours to know his subject. It is not enough to have the figures relating to a subject. If the statistician is to collate or derive statistics that are meaningful and useful, he needs to have a knowledge that goes beyond the mere figures. He must have a feeling for his raw material, and he cannot have this without such knowledge.

Secondly, he is a seeker of the truth, a researcher, a scientist. He has an analytical approach to his work, and is prepared to present what he believes to be the truth, both basic and inferential, without fear of the consequences.

Thirdly, he is ingenious and inventive. He is constantly searching his mind for improved methods and formulae, and is always prepared to question the validity and practicability of those currently accepted and the assumptions underlying them.

And finally, he is resourceful. He lives in pioneering days. To him statistical method does not consist of a set of theories or theoretical principles. It is an applied science. He brings to bear on his work a large measure of logic, common sense, and foresight.

One word more. In my work on *Planning Profit Strategies*, I demonstrated in Appendix 1 some of the more useful methods of regression analysis. In writing Chapter 7 of the present work, I drew on this to some extent, and found in so doing that my exposition of the method of successive elimination was not as clear as it might have been. I have put this right in Chapter 7 by a fuller extension of the example used for demonstrating the method.

My thanks are due to The Financial Times Ltd. for permission to use a real-life example of a decision tree; and to Mrs Margaret Gittins of Tewkesbury, for the painstaking way in which she prepared the typescript.

E. J. B.

Tewkesbury
June 1972

Acknowledgements

We are indebted to *The Financial Times* for permission to reproduce an extract from the article 'How to find Happiness with a Decision Tree' by David Palmer, *Financial Times*, 18th May, 1971.

1

Introduction

Management statistics form part of the management information of any progressive company—a part that has grown in relative importance very appreciably since the Second World War. There are several reasons for this. Marketing managers, for instance, are no longer satisfied with a general statement to the effect, say, that a particular market is highly competitive. They want a numerical measure of the degree of competition that their companies' products are meeting or are likely to meet. Hence, the rapid development of market research, which is largely an exercise in the collection and analysis of statistics. Similarly, the product cost accountant is now less interested in prime cost than in the product-variable or marginal cost. Another reason is that there is a growing recognition that business managers, like scientists, with a knowledge of statistics are better at their jobs than those without.

A CLASSIFICATION OF MANAGEMENT STATISTICS

For practical management purposes statistics may be divided into seven main categories, as follows:

Internal:		
Routine:	basic	derived
ad hoc:	basic	derived
External:	basic	derived
Internal–external:	–	derived

There are, of course, numerous subdivisions of all these categories. Internal routine statistics, for instance, both basic and derived, may be subdivided by reference to department of origin; the accountant's for financial statistics, the production manager's for production statistics, the marketing manager's for sales statistics and so on. External statistics, too, can be subdivided by reference to source: published, official and non-official; the market research office for external statistics, usually *ad hoc*; trade associations for collections for private distribution amongst member companies.

Basic statistics

Basic and derived statistics are two kinds of statistics which imply two kinds of statisticians, viz.: the statistical collator and the statistical analyst. The raw material of the former consists of returns, detailed books of account and other records. The raw material of the latter is the output of the former, i.e., statistical aggregates collated in conformity with some predetermined classification. A figure of total sales proceeds is a useful piece of information; but how much more useful if it is analysed into classes and subclasses by reference to products and markets?

It is sometimes said that the only true statistician is the statistical analyst. But this is scarcely true. Many series of our published basic statistics have called for much thought and expert knowledge on the part of the collators. An examination of the national accounts, which are published annually in the *National Income* blue book, with quarterly figures in *Economic Trends*, provides the evidence in support of this. The original compilers had to know a great deal about economics, accountancy, the principles of classification, and the requirements of statistical analysts. Above all, they had to have a thorough knowledge of their own sources of raw material and a thorough understanding of the definitions, and of the application and limitations of index numbers and other statistical devices. It is true to say that the statisticians responsible for compiling our national accounts are specialists in their field. Statistical collators generally, even more than statistical analysts, need to have a specialised knowledge of the field in which they are working.

Derived statistics

Averages, ratios and parameters are examples of derived statistics. Although some of them are exceedingly simple in derivation and definition, they represent the output or part of it of statistical analysts, who know what averages have significance for information and for analytical purposes, as well as knowing how to derive them. Derived statistics are inferential data, often subject to a margin of error. An average as such is worthless. Its value lies in the uses it can be put to; and it is when it is applied to some practical or theoretical purpose that it acquires a margin of error. It is in no way an object of this work to discuss the theory of statistics. Theoretical statisticians toss coins and throw dice to build up frequency distributions and formulate a theory of statistical probability. And while admittedly these exercises are not without practical value, they are liable to become an obsession with those who carry them out to the exclusion of all practical considerations. The mathematics are profoundly interesting and absorbing, and are best left to the academics with time to spare. There is another difficulty: the practical results of these theoretical and mathematical exercises are given expression in the form of a theory of error, which is necessarily based on certain assumptions about the nature of the basic data. Some of them are scarcely valid in the field of management and economic statistics.

Amongst the routine statistics maintained by most transport departments which operate fleets of road motor vehicles, is fuel consumption and vehicle mileage. Such statistics have a number of uses. They provide a check on engine efficiency and give warning to the repair engineer of the need for engine overhaul. They provide a check to thefts of fuel. And they are invaluable to the investment appraiser when vehicle renewals or fleet expansions are under planning consideration.

Now we may accept the fuel consumption and mileage of any vehicle for any particular week as accurate, so that the derived figure of miles per gallon for that vehicle in that week would be equally accurate. There would be no room for error. However, m.p.g. depends upon such factors as the maximum speed attained and length of time sustained, the number of starts from cold, the condition of the vehicle—its engine, carburettor or fuel injector, and brakes, and traffic conditions. The m.p.g. for the vehicle for the week's work is therefore not representative. From this point of view it does indeed contain a margin of error. What we need is a much larger sample, one covering a much longer period than a week, if we need a representative m.p.g. for the vehicle itself, and one embracing other vehicles of the same type, if we need a representative m.p.g. for all vehicles of the type. Once we have this representative m.p.g. and the calculated error margin, that is, the standard deviation for a 68 per cent probability and twice the standard deviation for a 95 per cent probability, we can then compare one week's performance of the vehicle or any one of the vehicles of the type with the representative m.p.g. If the performance falls within one standard deviation of the representative m.p.g., it may be regarded as satisfactory; if outside it, an investigation may be necessary; and if below the representative m.p.g. less twice the standard deviation, an investigation would be essential. A downward trend of the m.p.g. for the vehicle in the last ten or fifteen weeks may suggest engine deterioration. An examination of the trend should be the first step in an investigation.

It is in this direction that error theory comes into its practical own. Its basis is the normal frequency distribution or the normal curve of error. Wherever a distribution is normal or near normal, the standard deviation provides a measure of the error for a 68 per cent probability that any particular observation will fall within its limits. If the representative m.p.g. is 25 and the standard deviation 2, then we can express the m.p.g. as 25 ± 2. Of all the observations making up the representative m.p.g., 68 per cent of them would fall within the range 23–27.

There is good reason for supposing that the frequency distribution of m.p.g. figures is normal; but it can easily be tested. There is a formula for it called the chi-squared test; but if the mode is equal to the mean, the distribution is probably normal. Suppose the number of observations used for calculating the representative m.p.g. is 64 distributed as follows:

m.p.g.	19	20	21	22	23	24	25	26	27	28	29	30	31
Number of observations	1	2	3	5	7	9	10	9	7	5	3	2	1

This is a perfectly normal frequently distribution—of a perfection never, or hardly ever, to be found in practice. The mode is 25 m.p.g., that is, the figure that occurs most frequently; and the mean, or simple average, is also 25 m.p.g. Such a distribution is represented graphically with m.p.g. measured on the horizontal axis and the number of observations on the vertical axis. Where the distribution is normal, the fitted curve rises to the right slowly at first, then more quickly, reaching a peak at the mode where it begins to decline—the complete graph being bell-shaped and symmetrical, like Fig. 1.1. Figures 1.2. and 1.3 represent *skewed* distributions, where the mode is smaller than the average in the left-hand skew and greater in the right-hand skew.

The formula for calculating the standard deviation and the methods of taking account of two or more factors that can be measured numerically will be given in later chapters. In the case of vehicle fuel consumption, for instance, the number of cold starts is measurable, so too is the number of calls made, and the tonnage carried. The mechanical condition

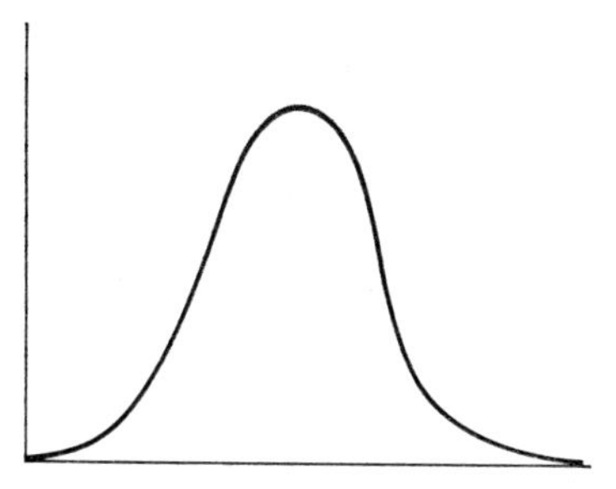

FIG. 1.1 Normal distribution

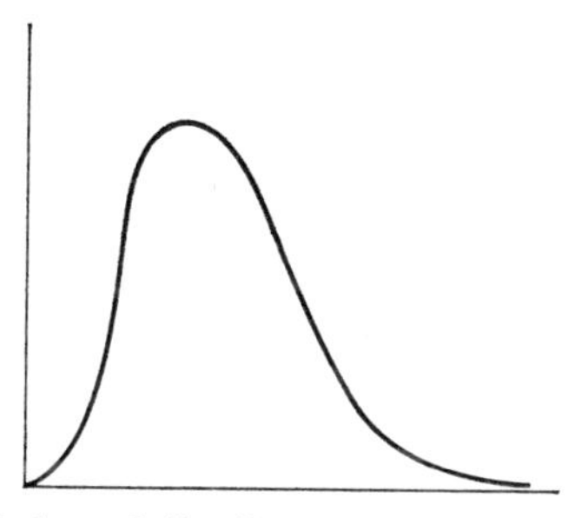

FIG. 1.2 Left-hand skewed distribution

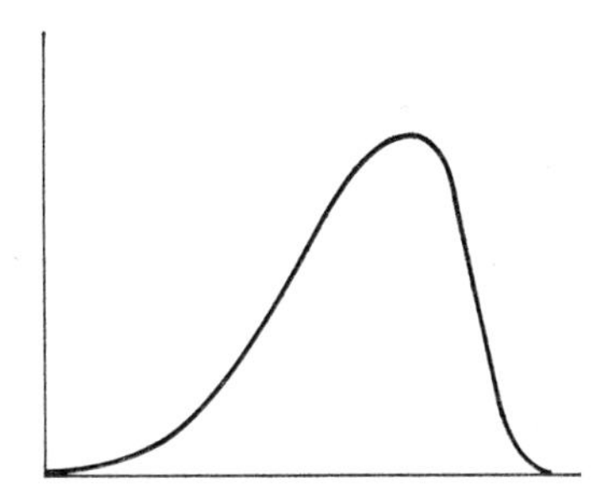

FIG. 1.3 Right-hand skewed distribution

of a vehicle may be measured in terms of mileage since the last engine overhaul. As to traffic conditions, it may be possible in some circumstances to obtain a measure of it from the statistics of m.p.g. in traffic. It would never be perfectly satisfactory; it would serve as an indicator, and its significance would certainly need to be tested. Of course, one could call it the average m.p.h. factor; the difficulty is that a vehicle that spent most of its time travelling at high speed on motorways may do much the same low m.p.g. as a vehicle that spent its time in town and urban areas. Factors of this kind are best left out of account either as factors in their own right or as indicators of other factors. In short, a sound hypothesis is essential in this as in other kinds of statistical analysis.

Statistical inference

Statistical inference is a term one sometimes hears used amongst statisticians, and it occasionally appears in the literature. It is something of a pretence, for it can mean nothing more than inference drawn from statistical data expressed in the form of premises. It is just as much the fruit of general logical reasoning as any other kind of inference. Whether the premises take the form of statistical data or not, no more can be extracted from them than has gone into them. An imprecise inference may be drawn from precise premises, but not a precise inference from imprecise premises. Nothing is more precise than a numerical evaluation. A precise inference drawn from imprecise premises is a *non sequitur*, of which the following is an example.

> This was equivalent to an increase in output per wage-earner of 47 per cent and, since the length of the normal working week had been reduced during the period, to an increase in output per man-hour of 65 per cent.

Surprisingly, this passage is quoted from a paper written by a professor of economics and published in a learned society's journal. Whether one could call the conclusion statistical inference or not, it is not borne out by the premises; it is too precise.

Although mathematical reasoning is a process apart, it contains an element of general reasoning. A person incapable of reasoning logically would make a poor mathematician. If $X=Y-2$, then $Y=X+2$ involves a process of reasoning. A mathematician would employ formal mathematical reasoning to arrive at the conclusion that $Y=X+2$, but the same conclusion can be reached by purely general reasoning. It could be argued that all formal mathematical reasoning is based initially on general reasoning, in so far as it was general reasoning that discovered the formal mathematical rules, for instance, that one can transfer a term in a linear equation from one side to the other provided one also changes its sign, which simply means that if one adds (or deducts) a quantity to (or from) one side of an equation, one must also add (or deduct) an equal amount to (or from) the other side. Mathematics is general reasoning conducted in symbols and numerical values, some symbols denoting hypothetical values; others, arithmetical processes. Statistical argument is general reasoning conducted in mathematical terms. Mathematicians never speak of *mathematical inference*.

It does not follow because the premises of an argument are precise, and the stated inference imprecise, that the inference is necessarily valid. The term *statistical inference* is often used in a sphere of statistical analysis where imprecise inferences—often invalid ones—are drawn from precise premises. The sphere is that of statistical correlation, which forms the subject of Chapter 5.

Questionnaires and collection

Simplicity is the keynote in the composition of a questionnaire of any kind. The questions asked should be relevant and seen to be relevant to the object of the enquiry, which should be briefly stated in an introductory note. The introductory note needs to be a short essay in persuasion. The more persuasive it is as to the rectitude and usefulness of the objects of the enquiry, the greater interest will recipients take in completing the form, and the greater their response.

Questions should be clearly and concisely stated. If possible they should be arranged in an order that helps the recipient. Legal terms and technical jargon are best avoided, even where it is known that all recipients will understand their meaning. This applies more particularly to external questionnaires, especially where recipients are to be a cross-section of the public at large, or are to be, say, the retail grocers in a test area.

It is desirable to keep the amount of clutter in questionnaires to a minimum, *clutter* being defined as space or material provided 'for office use only'. Where clutter is necessary, it should be located in the form in as inconspicuous a place as possible. Code numbers for electronic data processing purposes, for instance, are usually best printed in a separate column on the extreme right.

Size of paper and layout are important. It should be possible for a respondent to type the answers on a machine with a standard length carriage, which is 10 inches. If foolscap is used for the questionnaire, the layout should run down the length of the paper and not across it. Always consider the convenience of recipients in other ways. Enclose a prepaid return envelope; and never on any account enclose advertising and similar literature, which is a form of clutter.

Most questionnaires are sent to a sample of the total population. Methods of statistical sampling, the general technique of sampling, and related matters form the subject of Chapter 4. It is essential to preserve the representativeness of a carefully selected sample by obtaining a large percentage response, even where the sample itself is very large. Responders and non-responders are two distinct classes of a population, and neither is representative of the other, nor therefore of the whole population. A large proportion of non-responders creates bias; and it is useless to increase the size of sample with the object of allowing for a percentage of non-responders.

Any population, which in statistical terminology means the aggregate of a class of companies or people whose activities are subject to investigation by sampling, and a sample itself, can be divided into classes by reference to the response: class I consists of recipients of the questionnaire who respond immediately; class II, of those who respond after one reminder; class III, of those who respond after two reminders; class IV, of those who respond only to a personal visit; and finally, class V, of those who flatly refuse to co-operate at all. It is a prayer of all statisticians engaged in a sampling exercise of this kind that the proportion of flat refusals will be negligible. If the number of returns falling into each class is large enough, the results obtained from each will be representative of the class. Then a trend of bias may be established, and from this it may be possible to estimate the bias of the flat refusal class, and so obtain results representative of the population as a whole. It may be worth mentioning here that on the whole, investigation shows that the class I response averages half the total sample; the class II response, half the remainder; the class III response, again half the remainder; the class IV response (by telephoning and personal calls) accounts for a further response, which may be about half the total falling into classes IV and V, so that class V, the refusals, may account for as much as 6 per cent of the sample, which is not by any means insignificant. These figures are averages based on a wait averaging four to six weeks between the despatch of the questionnaire and the first reminder, between the first reminder and the second, and so on. It was found that recipients take longer to return a complex form than a simple one, complexity being measured in terms of the number of questions. There are, of course, other factors. A single question may take a day's research to find the answer, others only a few minutes.

Chasing or *rounding up* non-respondents as the process of sending reminders, telephoning and paying personal visits, is variously called, can be very expensive in clerical labour, travelling, stationery and postage. Any organisation that is thinking of embarking on a sampling investigation by questionnaire must be prepared to meet the cost of a routine set-up for the purpose of chasing. It is a cost that is liable to be overlooked at the project planning stage.

Investigation by internal questionnaire is an entirely different matter. The number of companies that have branches and other establishments and agencies spread over the country has grown rapidly in recent years. *Ad hoc* enquiries are sometimes necessary; but routine regular weekly, monthly and annual financial and statistical returns sent by branches to headquarters, which form an essential part of the control system, should be wide-ranging enough for most purposes, and designed to do away with the need for *ad hoc* enquiries. Whenever an internal *ad hoc* enquiry is thought to be necessary, the routine system of returns should be reviewed. Of course, it sometimes happens that an *ad hoc* enquiry by questionnaire is sometimes necessitated by a need to discover whether an additional routine return would be worthwhile.

O & M is usually responsible for the design, content, layout, coverage and wording of all forms, and it provides a convenient central co-ordinating body for overseeing all standard forms and questionnaires used in the organisation. One of its major tasks as a co-ordinating body would be to reduce overlapping to a minimum. It would then be impossible for an internal *ad hoc* questionnaire to ask questions which are already answered in the routine returns.

It is the market and economic research offices of companies that tend to place dependence on internal as well as external *ad hoc* questionnaires for the information they need. Some of the data they seek may be already available in the office in detailed and collated form. It is important that waste of this kind as of any other kind should be prevented; and the best way of doing it is to compel the research office to conform to the rules regarding the submission of proposed internal questionnaires to O & M for approval. In its own interests, it should at least consult O & M on the design of all questionnaires both internal and external.

2

Tabular and diagrammatic presentation

How should statistics be presented in an article, report or paper? In tables or in diagrams? To a great extent, the answer depends upon the potential readers. Statisticians, accountants, research economists and other people accustomed to handling numerical data, probably prefer tables of figures to diagrams. Figures are more precise than readings from a diagram, and the reader can the more easily apply them to his own purposes. In this chapter, we are concerned with presentation only. Statistical research often involves the use of functional diagrams, which form a subject of Chapter 5. A functional diagram is one that helps the investigator to determine the kind of mathematical empirical relationship that exists between two or more factors. It provides a visual aid to an understanding of the relationship, an understanding which tabulated data alone cannot give.

STATISTICAL TABLES

There are two kinds of statistical table, one designed for information, and the other for the purpose of bringing out the point of the argument of a chapter, report or paper. Order and classification are essential features of any statistical table. Apart from this, the form of a table made for information is not important, though it has been argued with some justification that since a time series is easier to read down a column than along a line, the periods or dates should provide the line headings (or *stub*) and not the column headings. There are exceptions to this rule, if such it can be called, for one or both of two good reasons. One is that the reader may be expected in some cases to be more interested in comparing one item with another rather than with the movements of each item in time. And the other is that items that cannot be shortly defined are more easily and economically accommodated in the stub than in column heads. It is probable that both had something to do with the Central Statistical Office's setting out all time series in *The National Income and Expenditure* with items shown in the stub and periods in the column heads; though admittedly accountancy convention may have been the deciding factor.

Company accounts do not set out to give time series. The published accounts give the figures for two consecutive years, and though investment analysts and others undoubtedly compare one year with the other, the main interest lies in the absolute level of the figures for the later of the two years. Many public quoted companies now give in their published accounts a summary of the principal figures for the last ten years or so sometimes including one or two ratios such as earnings to dividends (the *times covered* in stock market jargon). Practice varies somewhat in the matter of layout. Nearly all those that give a summary show the years in the column heads and the item definitions in the stub; the years run from left to right, the latest year being in the last column. A few present the series beginning with the latest year and ending with the earliest. Only a very few companies conform to the 'rule' and show the years in the stub with the item definitions in the column head.

Official practice is to present time series to read downward, i.e., with periods and dates shown in the stub. Every time-series in *Financial Statistics*, for instance, is presented in this way; and most of those in the *Monthly Digest of Statistics* conform to the rule, the exceptions being where the items have exceptionally long definitions. Oddly enough, only a few of the time series published in the *Annual Abstract of Statistics* conform to the rule, nearly all read across, with years given in the column heads. Most other official publications conform to the rule: the summary tables in the Board of Trade *Business Monitor*, and the regular time series in *Economic Trends*, for instance. On the whole, the Central Statistical Office's and other Government departments' practice appears to favour the rule rather than the reverse, whereas company practice in published accounts does not. The fact remains that it is much easier to follow the ups and downs of a time series when it is presented vertically than when it is presented horizontally.

Table twisting

Not all tabulated statistical data consist of time series. The art of table twisting is an accomplishment worth practising where the data are designed to emphasise or demonstrate the point or strength of an argument. It is more generally applicable to non-time series, though it is not without value for time series too, especially where the run of periods is short. Suppose it is necessary to tabulate the purchases of ordinary shares by insurance companies quarter by quarter and year by year for the five years 1965–9. The figures for the UK are given in a single column in a supplementary table in *Financial Statistics* for April 1970:

	Quarterly (£m.)	*Annual total (£m.)*
1965: 1st quarter	53	—
2nd quarter	50	—
3rd quarter	43	—
4th quarter	60	206
1966: 1st quarter	66	—
.	.	
.	.	
.	.	
.	.	

To complete the table in this form would cause an unnecessary waste of space. The space-saving alternative presentations are:

First alternative layout (£m.)

Quarter	*1965*	*1966*	*1967*	*1968*	*1969*
1	53	66	45	122	120
2	50	60	69	152	150
3	43	48	82	158	112
4	60	47	117	163	138
Total (for year)	206	221	313	595	520

Second alternative layout (£m.)

	1st qtr	*2nd qtr*	*3rd qtr*	*4th qtr*	*Annual total*
1965	53	50	43	60	206
1966	66	60	48	47	221
1967	45	69	82	117	313
1968	122	152	158	163	595
1969	120	150	112	138	520

This is a perfect example of table twisting. Both consist of five columns and five lines, so that there is nothing to choose between them from the space-saving point of view. The former emphasises the quarterly trend in each year, whereas the latter brings out the comparison of each quarter with the corresponding quarter in other years, which is important where there are significant seasonal variations. Consider the central government's net balance or, as it is now more generally called, *the borrowing requirement.* Owing to the incidence of receipts from taxes on income, seasonal variations are appreciable, and of the two alternative layouts showing quarterly figures for a short run of years, the latter of the two forms above is the more appropriate:

Government net balance (£m.)*

	1st qtr	*2nd qtr*	*3rd qtr*	*4th qtr*
1966	−833	438	424	492
1967	−627	423	542	796
1968	−426	394	360	427
1969	−1,454	234	85	481

* Surplus (−).
Source: Financial Statistics.

A table in this form provides a simple means of what, in its effect on the eye of the reader, amounts to the removal of seasonal variations. As the eye roves down each column, it is comparing like season with like season.

Effective table twisting for non-time series has much the same kind of considerations to be taken care of. Non-time series mostly apply to a period of time or some particular date, which should generally be stated. Some numerical data are timeless universal truths; they are rarely if ever statistics in the accepted sense of the term. Examples are the ratio of the circumference of a circle to its diameter ($\pi = 3{\cdot}1416$), the base of natural logarithms (epsilon$=2{\cdot}7183$), the velocity of light (3×10^5 kilometres per second) and absolute zero temperature ($-273{\cdot}2$°C). Some apparently timeless constants are really statistical averages based on data for long periods of time, e.g. life expectancy for healthy males or females of a given age, and average annual rainfall.

There is a table in *Financial Statistics* headed 'Sector capital accounts'.

It consists of six tables, each of which relates to a sector of the economy. The statistics given consist of time series under five heads for the years 1966 to 1969 and for the quarters in each of the years 1967 to 1969. Suppose we are interested only in the 1969 figures, and we wish to present them to the best advantage. Table 2.1 presents the figures in one way, and Table 2.2 presents the 'twisted' version. Probably the better presentation for most purposes is that of Table 2.1. But if we wish to compare the several kinds under each sector separately, then Table 2.2 provides the better presentation.

Now suppose we wish to compare two years' figures throughout; we would then have a third dimension, and a necessary doubling either of the number of columns or of the number of lines. There are, then, now four ways of presenting the figures, the *pro formas* being as shown in Tables 2.1(*a*) and (*b*), and 2.2(*a*) and (*b*).

TABLE 2.1 Sector capital accounts, 1969 (£m.)
First alternative

	Saving (*1*)	*Capital transfers (net receipts)* (*2*)	*Gross domestic fixed capital formation* (*3*)	*Increase in stocks and work in progress* (*4*)	*Net acquisition of financial assets** (*5*)
A Central government	2,902	−1,590	488	40	784
B Local authorities	471	116	1,738	—	−1,151
C Public corporations	842	1,326	1,487	−10	691
D Financial companies	324	−53	466	—	−195
E Industrial companies	2,921	585	2,722	946	−162
F Personal sector	2,370	−384	1,052	188	746

* Col. (5) = (1) + (2) − (3) − (4).

For most purposes, where the initial object is to compare the two years, Tables 2.1(*b*) and 2.2(*b*) are the best. The former of these reflects the tabulated form of the published table.

Quarterly figures for the two years could be introduced either as part of the third dimension or as a fourth dimension. The (*b*) formulae of Table 2.1 or 2.2 would be best for the former, and the (*a*) formulae

TABLE 2.2 Sector capital accounts, 1969 (£m.)

Second alternative—the 'twisted' version

	Central government (1)	*Local authorities* (2)	*Public corporations* (3)	*Financial companies* (4)	*Industrial companies* (5)	*Private sector* (6)
A Saving	2,902	471	842	324	2,921	2,370
B Capital transfers (net receipts)	−1,590	116	1,326	−53	585	−384
C Gross domestic fixed capital formation	488	1,738	1,487	466	2,722	1,052
D Increase in stocks and w.i.p.	40	—	−10	—	946	188
E Net acquisition of financial assets*	784	−1,151	691	−195	−162	746

* See footnote to Table 2.1.

TABLE 2.1 (*a*) Years accommodated in column heads

	Savings		*Capital transfers*		
	1966	*1969*	*1966*	*1969*	......
Central government					
Local authorities					
....					

TABLE 2.1 (*b*) Years accommodated in the stub

		Saving	*Capital transfers*	....
Central government	1966			
	1969			
Local authorities	1966			
	1969			
.				
.				
.				

TABLE 2.2 (*a*) Years accommodated in column heads

	Central government		*Local authorities*		
	1966	*1969*	*1966*	*1969*	....
Saving					
Capital transfers					
.					
.					
.					

TABLE 2.2 (*b*) Years accommodated in the stub

		Central government	*Local authorities*	....
Saving	1966			
	1969			
Capital transfers	1966			
	1969			
.				
.				
.				

for the latter, for which the following is the appropriate *pro forma* based on Table 2.1(*a*).

	Saving		
	1966	*1969*	....
Central government			
1st quarter			
2nd quarter			
3rd quarter			
4th quarter			
Local authorities			
1st quarter			
·			
·			
·			

Needless to say, table twisting is a preliminary process, mostly carried out before a single figure is posted. Figures have to be posted sometimes however, when the statistician feels unable to make a choice from his *pro formas*. With the figures of some of them entered, a sound decision may be easier to come by.

It is worth mentioning that the space occupied by each single figure in a statistical table is usually referred to as a cell. A cell can be located in the same way as a place on a map. Cell C(3) in Table 2.1 above is occupied by the figure £1,487 m. and B(3) in Table 2.2 by £1.326 m. A comparison of the cells of the two tables shows that A(1) in both contains the same figure, so also does the last cell, at the extreme bottom right hand corner. All other cells have changed: A(2) in Table 2.1 becomes B(1) in Table 2.2, A(3) becomes C(1), and so on. Table twisting, from this point of view, becomes a matter of changing the pattern of cells from one systematic form to another. Statistical tabulation is the transformation of a potential chaos of data to order and classification. It would be possible, but not quite so easy, and certainly less effective, to present any statistical data in a chaotic form. Consider the data given in Tables 2.1 and 2.2 for instance. It could read something like this:

	£m.
Capital transfers of public corporations	1326
Saving of financial companies	324
Increase in stocks of industrial companies	946
Net acquisition of financial assets of central government	784
...	

and so on until all 30 of the cells were exhausted. The chaos would contain the whole of the data, but to what purpose? Comparison would be almost impossible; and it would never be easy to find any particular figure.

STATISTICAL DIAGRAMS

Of the three main types of diagram in common use, the time chart is the most useful, largely because it presents a picture that cannot so readily be visualised from the figures themselves. The other types are bar charts and cake diagrams, both of use in different applications.

Time charts

A time chart traces the course of a series of statistics over a period of weeks, months, quarters, years or even decades, the course being mapped by a series of dots plotted on graph paper, and connected by straight lines. The lines connecting any dot with its neighbour should always be straight. There is a strong temptation amongst beginners to attempt to smooth out the pointed peaks and troughs to no useful purpose. It presupposes a knowledge of the position of hypothetical intermediate dots which cannot exist. Smoothing is a process designed to ascertain relatively long-term trends, usually presented on the chart as a straight line or sometimes a curve running continuously the length of the chart.

The time scale

Time is always measured on the horizontal axis or base line, and the charted factor or variable on the vertical. Each period of equal duration should be given equal length on the axis. Generally, with calendar monthly data, there is not much point in trying to adjust the length to accord with the number of days, or working days, in the month. Some organisations and industries base their short-period statistics on periods each of four weeks with 13 periods to the year. This avoids the difficulty caused by the varying length of the calendar month. But the system has its problems: once every seven years or so, it is necessary to drop a week to keep the periods in alignment one year with another; and it is not possible to build up quarterly statistics consisting as they do of 13 weeks, from four-week totals; the UK motor industry uses four- and five-week periods to solve this problem, there being two four-week periods and one five-week period to each quarter.

Although equal periods should be given equal space, it is the vertical lines of the graph paper that represent the periods in the best practice, not the intervening spaces. Figure 2.1 is the design for presenting annual figures; and Fig. 2.2 the design for presenting shorter-term figures for a period of years.

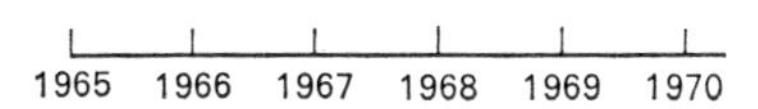

FIG. 2.1 Time scale for annual data

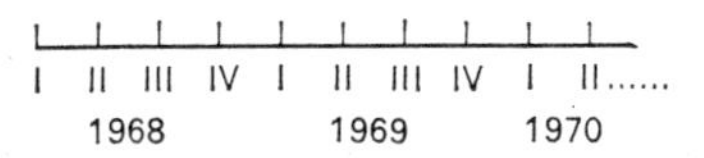

FIG. 2.2 Time scale for quarterly data

It does not matter whether the statistics relate to flows, like costs and revenues, or to points of time, like debtors and creditors, stocks, and other balance-sheet items, these are undoubtedly the best forms of calibration and marking of time scales. To insert the years in Fig. 2.1 between the lines is ambiguous and often very puzzling. Some published time charts show each plotted dot for flow halfway between the consecutive lines: the draughtsman often has to draw a pencilled line halfway between the permanent uprights to make sure he is inserting the dot in the correct position.

The time scales demonstrated in Figs. 2.1 and 2.2 are continuous, i.e., the years follow on one after another along the length of the chart. There is another form of time-chart, which can conveniently be called seasonal. It applies to short-period statistics, but could be adapted to annual statistics in quinquennial periods or decades for the analysis of cyclical movements. For the time seasonal chart, the time scale covers one year only. The month of January or the first week or quarter is indicated by the first mark on the scale to the right of the vertical axis, which although not marked as such represents the month of December or the last week or quarter of the previous year as the case may be. Each year's figures are then plotted on the same scale, and the line of each is appropriately marked with the year, or alternatively a key is given in a corner indicating the colour or style (dotted, continuous, discontinuous and so on) of the line for each year. Figure 2.3 provides an example of a quarterly series plotted on a seasonal chart.

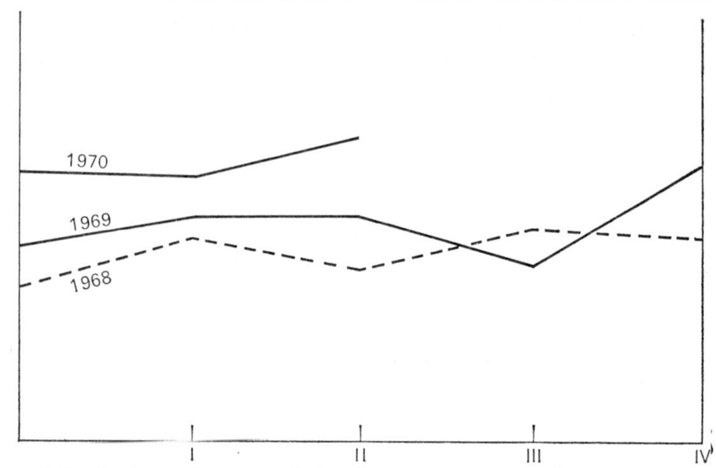

FIG. 2.3 A quarterly seasonal chart

Observe how the graphs for 1969 and 1970 begin on the vertical axis at the same levels as the terminating levels of 1968 and 1969 respectively. This device lends a measure of continuity to the plotted series which it would otherwise lack.

Of the advantages of the seasonal chart, the principal one is that it never needs extending to the right. It may have to be extended upward where there is a general upward trend from year to year; but it should not be difficult to make some allowance for general trend when the diagram is designed. Theoretically, a seasonal chart could be used indefinitely; but in practice it is limited to the presentation of a few years' graphs only, the number depending upon the dimensions of the chart, the skill of the draughtsman and the relative levels of the several years' graphs. Another advantage is that it gives an instant picture of the seasonal pattern, which a continuous chart does not.

Although it is much easier to read and compare a vertical array of figures in a column than a horizontal array in a line, the mind revolts at the idea of a vertical graph of a time series with the time scale given on the vertical axis, in which up and down becomes side to side. Undoubtedly, we could accustom our eyes and minds to the idea that a movement to the right is upward, and one to the left is downward, just as we have accustomed our minds to thinking in terms of new pence, and may have to later in terms of metres and centimetres, grammes and kilogrammes, litres and kilolitres and ares and hectares. The horizontal time scale for charting purposes is both sensible and conventional.

The vertical scale

As with the horizontal time scale, equal values measured on the vertical scale should be given equal length in the calibration. There is one exception to this rule and that is where a logarithmic scale is used, a form of presentation that is considered below. The form of presentation that gives equal length on the vertical axis to equal quantities or values is referred to as the *rectilinear* or the *arithmetic* to distinguish it from the logarithmic. A logarithmic scale is sometimes called a *ratio scale*, because it gives equal lengths to equal ratios: no matter what may be the absolute value of, say, an increase of 10 per cent, it is given a length on the scale equal to that given to an increase of 10 per cent at any other level of absolute value.

The arithmetic vertical. In the simplest form of time chart, the vertical axis begins with zero at the intersection with the horizontal axis. In practice, it is rarely possible to have a zero start without wasting a great deal of space. Indeed, if the range between the highest and lowest readings of a time series is very small, say a matter of 5 per cent or less, the plotted graph may present itself to the eye as a horizontal straight line. Sometimes, too, statistical time series consist partly of negative figures and partly of positive figures: loss is negative profit and profit is negative loss. Negative and positive figures are found together where a series represents the difference between two magnitudes such as annual cost and annual revenue, or where it represents changes in some magnitude, such as the money supply.

It follows that there can be no conventional starting value for a vertical scale in a time chart. If the graph is to serve its purpose of presenting to the eye a series of figures having a narrow range, the starting value must be greater than zero, and where there are negative values to be accommodated, the starting value must be less than zero. In analytical graphing, which forms a subject of Chapter 5, the axes may form a cross intersecting at zero for both factors involved, each having two negative and two positive quarters in the complete diagram. For time series, however, the two axes so called, are not strictly axes in the mathematical sense, they are merely a means of displaying the calibrations or scale used. The horizontal scale line is better called the base line.

In graphing a narrow-range series of statistics, the first step is to ascertain the range. The monthly average price index of 500 ordinary industrial shares of *The Financial Times*—Actuaries series ranged in 1969 between 146 in October and 189 in January. The range is $189-146=43$, which has to be accommodated on the vertical scale.

Most printed rectilinear graph paper is heavily lined at 1-inch intervals, and lightly lined at intervals of $\frac{1}{10}$ inch. The more nearly square the paper is, the more useful it is. In practice, the effective range is, say, $190-140=50$. If the graph paper has a depth of 10 inches, then allowing two-tenths to one point of the index would have the effect of using the paper to full advantage. The scale would start at 140 level with the base line and rise to 190. It could be marked at five-point intervals, i.e. every inch, or ten-point intervals, i.e., every 2 inches.

Suppose the chart is required to accommodate the figures for 1970 as well. Then some allowance must be made for the possibility that the range would be extended. As it happens, the figure of 189 for January 1969 is the highest recorded monthly average. At the time of writing there seems little prospect that this record will be broken in 1971. Nevertheless, there would be no harm in marking 200 the maximum on the vertical calibration. But the stock market has been falling since early April, and today, the price index stands at 132, after touching 130 a week ago. If the minimum monthly index number for the calibration is made 100 inserted level with the base line, the scale would have a range of 100. In that case, each ten points of the index would be given one inch on the scale. Alternatively, since the recent trend of stock market prices has been steeply downward, it may be thought wise to make the minimum 90 and the maximum 190, still leaving a full range of 100 points to 10 inches on the vertical scale.

The logarithmic vertical. As stated above, a logarithmic scale is a ratio scale. It can be made in either of two ways: by marking off the logarithms of the numbers on the line, but inserting the natural numbers, or by giving equal length to equal ratios on the scale. Printed log paper is available, but for a time series the type required is single log paper; for never on any non-analytical account should a log scale be used for measuring time along the base line of a time chart. Log-log paper or double log paper is limited in its use to statistical analysis.

Probably few statisticians use log paper or log-log paper, largely because printed scales rarely meet their requirements. Printed log scales always begin with a minimum of one natural. A single cycle scale rises to 10, a two-cycle scale to 100, a three-cycle scale to 1,000, and so on. If a single cycle scale takes 6 inches of line; then a two-cycle scale takes 12 inches; and a three-cycle scale, 18 inches. A log scale can never go down to zero natural, for the simple reason that log 0 equals minus infinity. The log of 1 equals 0, and that in effect is the starting point of the scale on log paper.

Figure 2.4 compares two log scales, one marked in logarithms and the other in the corresponding natural numbers. The scales extend over a single cycle. It will be seen at once that printed log paper has its limitations. Of course, 1·0 natural can be a unit of any size: 0·001, for instance, when the other end of the scale becomes 0·01, or 1,000, when the other end becomes 10,000. Any series of figures with a range in which the highest figure does not exceed ten times its lowest can with a little ingenuity be plotted against a single-cycle scale. The problem is

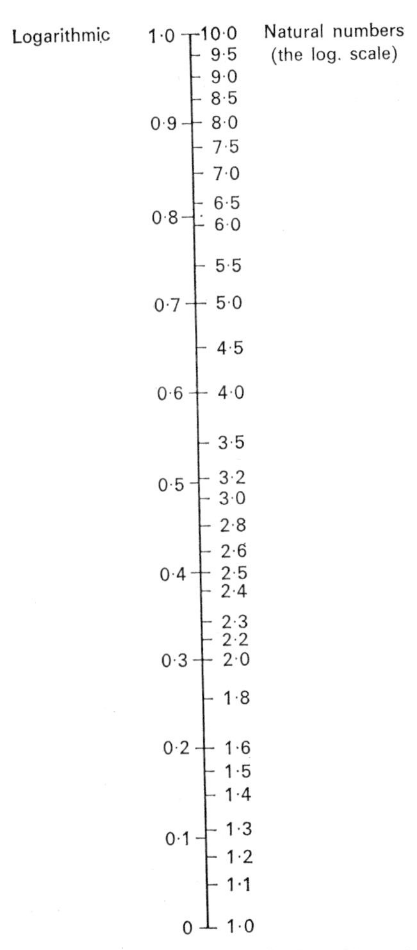

FIG. 2.4 A logarithmic scale, single cycle, showing the corresponding logarithms

that many time series have much narrower ranges than this, when a great deal of the available space is wasted. And some time series have a somewhat wider range, when two-cycle paper could be used, but again usually entailing much waste.

When statisticians think a ratio scale would be worth while, they usually convert the time series of statistics to logarithms and plot the logarithms on rectilinear paper. The logs can be inscribed in pencil for plotting purposes, and a scale of natural numbers substituted for them to produce a presentable chart. By this means, waste space can be reduced to a minimum. Consider, for example, the share price index numbers referred to above: the highest is 189 points and the lowest 146. To plot the series against a single-cycle scale would need only 13 per cent of the whole scale, 87 per cent being wasted.

For plotting the logarithms on a linear scale, we have:

$$\log 189 = 2{\cdot}2765$$
$$\log 146 = 2{\cdot}1644$$

$$\text{Range} = 0{\cdot}1121$$

Say, for charting purposes 0·12

by allocating $\frac{1}{2}$ inch to each 0·01 of a point 6 inches of the scale would be required, so that for 10 inch graph paper, 60 per cent of the scale would be used.

However, it seems that the use of log paper or logarithms for plotting time series has largely fallen into disuse. One sees a published table of the kind only very rarely nowadays, and then one wonders why such a device has been used. It hardly ever appears to serve any useful purpose. Where it might be useful is when a series is rising or tending to rise in geometric progression: a series rising regularly at, say, 5 per cent a month or year, would describe an upward turning curve on rectilinear paper; on log paper, it would describe a straight line. But generally, this would serve an analytical purpose rather than one of presentation.

Bar charts

Bar charts are the form of statistical diagram appropriate to the presentation of comparative statistics showing the constituent parts of some phenomenon at different places at the same time, or at different but widely separated times at the same place. They should never be used for presenting a continuous time series, except possibly where the object is to bring out changes in the proportions of constituent parts rather than to show their trends.

Suppose it is intended to show by bar chart the net capital stock by sector and type of asset for 1968. The first question to answer is whether each bar should apply to a sector of the economy or to the type of asset. The statistics are given in Table 2.3. Figure 2.5 contains a two-bar

TABLE 2.3 Net Capital Stock at Current Replacement Cost, 1968 (£ thousand m.)

	Vehicles, plant and machinery	*Buildings*	*All fixed assets*
Personal sector	1·9	18·7	20·6
Companies	18·0	12·9	30·9
Public corporations	10·9	5·9	16·8
Central government	0·4	3·2	3·6
Local authorities	0·9	20·5	21·4
TOTAL	32·1	61·2	93·3

Source: National Income and Expenditure, HMSO.

chart representing the distribution over sectors of each group of fixed assets. A five-bar chart could be drawn to represent the distribution over assets of each sector. The key is given in black and white; but for a wall chart, the polychromatic art of the draughtsman may well be given full play.

Cake diagrams

A typical cake diagram or pie chart is shown in Fig. 2.6. Its object is much the same as a bar chart without the element of comparison. For instance one could show by cake chart the distribution of all fixed assets in Table 2.3, or of vehicles, etc., or of buildings. Cake charts are used extensively for showing the share or slice of the whole cake received, or owned or enjoyed by each partaker. The shares of the total domestic product spent by consumers, public authorities and going into capital formation and stocks, or the shares of a market enjoyed by different producers, are examples which lend themselves to cake diagrams.

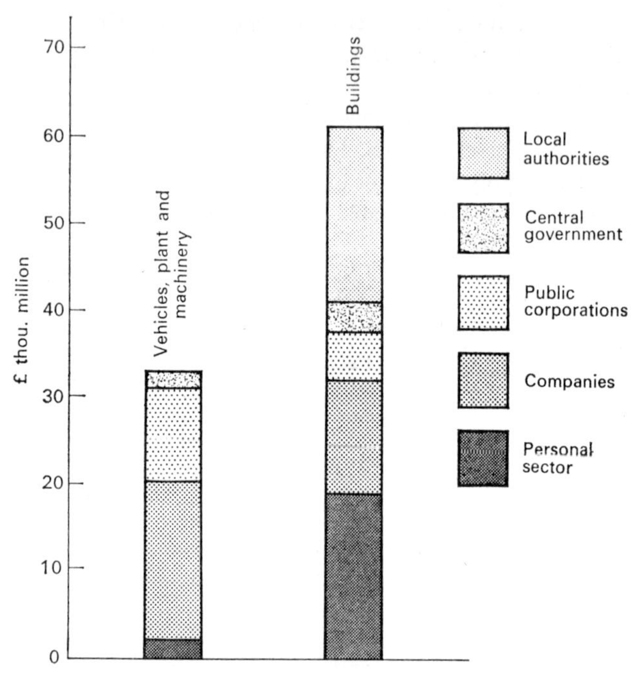

FIG. 2.5 Bar chart (based on Table 2.3)

Making a cake diagram is a simple matter. The slice of a cake is measured in any of three ways: on the arc of the perimeter, by the area, or by the angle described by the two radii which together with the arc circumscribe the slice: all bear the same relation to the total, i.e.:

$$\frac{\text{arc}}{\text{circumference}}=\frac{\text{surface area of slice}}{\text{surface area of cake}}=\frac{\text{angle in degrees}}{360^\circ}$$

Clearly, of these, the only one that can be used for making a cake diagram is the angle described by the two radii at the centre of the cake, so that the first step is to convert the basic statistics to degrees of the circle. Take, for instance, the total column of Table 2.3. The total of all slices is 93·3, which is represented by 360°. The slice owned by the personal sector is 20·6, which in terms of degrees of the circle is:

$$20{\cdot}6\times\frac{360}{93{\cdot}3}=79^\circ$$

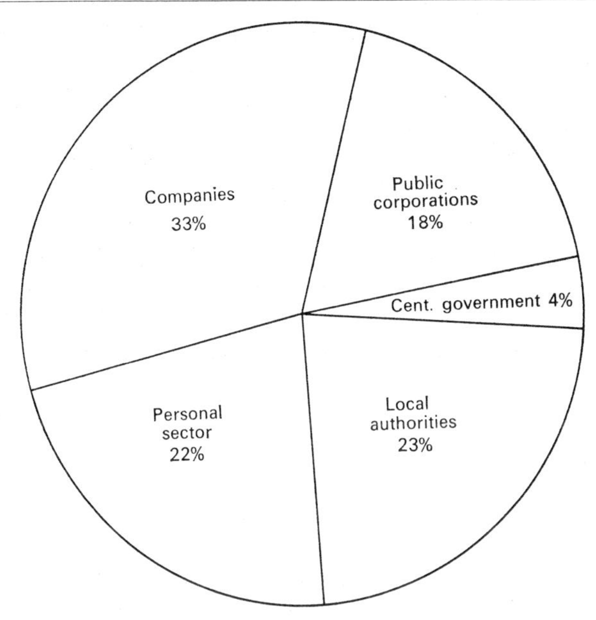

FIG. 2.6 Cake diagram or pie chart (based on totals in last column of Table 2.3)

Similarly with the other sectors. The total of all slices in terms of degrees is 360, which serves as a check on the arithmetic.

It seems to be the general practice to draw a radius down to the foot of the circle to serve as the first cut of the cake. Working clockwise, the appropriate angle from this is extended by means of a protractor to give the second cut and the first slice. Then with the aid of the protractor again, the next angle is extended from this second cut to give the third cut and the second slice, and so on. The first cut becomes the last cut, which provides a check on the geometry.

Cake diagrams are of no use at all where the value of the whole cake is unknown since in a cake diagram each slice represents not an absolute value but a proportional value. It is true that the absolute values may be written into the diagram: nevertheless, the object of the cake diagram is to represent proportions. Where the value of the whole cake is known, but the values of some of the individual slices are not, then the 'rag-bag' device which runs all through statistical classification can be employed. *Rag-bag* is statisticians' jargon for *residue*, and is usually written as *other* or *other types* or some similar residual wording. The values known collectively but not individually are given a single slice of the cake which is labelled 'other'.

Many statisticians appear to be coming round to the view that cake diagrams are not easy to interpret: to some eyes a slice of, say, 15 degrees on the NE side of the cake may appear larger or smaller than a slice of the same size elsewhere. An alternative, which has a number of advantages (a knowledge of the total value is not necessary, it is less deceptive, it is easier to grasp amongst others) is the single-entity bar-chart. It is similar to the bar-chart described and demonstrated above, except that each bar is undivided and represents a single entity. Each is proportional in length to the others, in accordance with the value it represents.

Pictorial diagrams

Apart from demonstrating the skill of the draughtsman, pictorial diagrams serve no useful purpose. Many are rather silly, especially those that represent the tonnage of various navies, population, airforce strengths and the like. They often show, say, 10 or 15 ships, men or aeroplanes plus half of one. One wonders whether there is any significance in the choice of the bows half of a ship in preference to the stern half; and half a man standing on one leg looks as though he needs a crutch. It calls to mind the efficiency auditor who is faced with a claimed saving of half a man: 'What half?' he asks.[1]

Of course, cake diagrams or pie charts could be presented pictorially. Cherries could represent some particular constituent, such as the tonnage of battle cruisers; slices of angelica, another constituent; and currants, yet another. But it is all very imprecise, unscientific and often quite silly.

[1] P. Philo in an article on 'The Savings Game' (*Management Accounting* of the USA, February 1970), pokes fun at some of the kinds of savings often claimed by departmental managers including the half-man kind. Pictorial charts are equally amusing.

3

Averages as types

It has been said that statistics is the science of averages. It is true that most, perhaps all, derived statistics could be logically described as averages. But the definition fails to take account of classification, collection, collation and presentation. Statistical analysis could more reasonably be defined as the science of averages. In this chapter we are concerned with the simplest kinds of averages, simplest in the sense that they are the easiest to derive. A classification of these simplest of averages is as follows:

- Arithmetic mean (AM):
  - Unweighted,
  - Weighted,
  - Moving average;
- Geometric mean (GM):
- Harmonic mean (HM):
  - Weighted
- Median
- Mode.

Of these, the most useful for statistical analysis is the arithmetic mean. The geometric mean has some serious limitations, although it was used extensively up to 1939 for making price and other index numbers. The harmonic mean is an inverted form, so to speak, of the arithmetic mean and has a few uses in statistical analysis. The median and the mode are statisticians' averages.

THE ARITHMETIC MEAN

The average of 3, 4 and 5 is 4. It is derived as follows: $3+4+5=12$ divided by 3, 3 being the number of items. Similarly, the average of

16 and 18 is 17:

$$\frac{16+18}{2}=17$$

Both these are simple averages or, in other words, unweighted arithmetic means.

Suppose a firm producing expanded metal has five machines all producing material of the same specification. For cost control purposes, the average output per machine is compared month by month. The following are the statistics for two consecutive months:

Machine	*Output in tons* *September*	*October*
A	24	24
B	26	25
C	25	—
D	23	21
E	40	44
Total	138	114
Machines in use	5	4
Average per machine in use	27·6	28·5

Machines A, B, C and D are of standard capacity, Machine C breaks down at the end of September and is out of service throughout October. Machine E is of double the standard capacity. It is clear that the increase in the average output is largely or entirely due to the idleness in October of machine C, and the continued use of machine E. Is it possible to eliminate the effect of this factor? The way is simple: it is to count machine E as two equivalent machines of standard capacity. We then have six equivalent machines in use in September and five in October, giving averages per equivalent machine in use of 23·0 and 22·8 respectively; so that far from there being an increase in machine efficiency or production in October, there was a slight fall. These are a kind of weighted average; the weighting has been carried out by adjusting the number of items, i.e., the denominator. In the common kind of weighted average, a system of weights is adopted, and is adhered to in all periods.

The weighted arithmetic mean

When statisticians speak of a weighted average, they almost invariably refer to the weighted arithmetic mean unless they qualify it. Indeed, any average unqualified may always be taken to be the arithmetic mean; although it may happen in a few contexts that the harmonic mean is intended.

Weighted averages are usually more accurate, where they are applicable, than unweighted averages, in the sense that they are more nearly representative. They are used in averaging averages and ratios. The admitted need for weighting amounts to an admission that averages and ratios are of varying significance, indeed, that a change in an average or ratio may be of no significance at all. The average number of cars owned in a village may double overnight, so to speak, from, say, 0·1 to 0·2 per household. If there are ten households, then the doubling of the average would arise from an increase in the number of cars owned from one to two. Neither of the averages is of any significance. Nor therefore is the ratio of one to the other, especially as an increase in the number of cars owned in the village is bound to give a ratio of at least two. But a ratio of two for a country or city would be significant indeed, even if it took 12 months or more to achieve it.

An increase in the weekly average consumption of tea per household in the village from 3 oz. to 6 oz. over a period would be rather less insignificant; so, too, would the ratio of $6/3 = 2$. For one thing, the total quantity of tea consumed by the villagers may vary by infinitesimally small quantities, whereas the number of cars owned can increase only by one car or more at a time. For another, tea is consumed by all or nearly all households in the village, and a change calls for no special effort by any household.

Statisticians have names for these two kinds of figures. A set of statistics expressed in terms of number where each unit is indivisible, is called a *discrete series*; and a set of figures expressed in terms of weight, value, volume, area or length, is called a *continuous series*. Oranges in terms of numbers give a discrete series, whereas oranges in terms of weight or value give a continuous series. Liquids and commodities like grain, flour, sugar and salt, give continuous series until they are packaged, when they can become numbers and give discrete series: pint-bottles of milk, cwt. sacks of grain and so on. The distinction between *discrete* and *continuous* should not be taken very seriously. Many liquids and similar commodities are delivered in bulk in containers of a standard size. To know the gallonage or tonnage is to know the number of barrels, tanks, sacks or cases, so that a set of figures of deliveries in terms of volume or weight can often be regarded as discrete rather than continuous.

However, in most cases, the distinction is never in doubt. It may not be possible to buy tea at retail in quantities smaller than a quarter of a pound, but this does not mean that a household or village must consume an exact number of quarters of tea in a week. It may consume any quantity from a teaspoonful upward. With cars owned outright, the number is necessarily exact: it may be one or two or three and so on; it cannot be $\frac{1}{2}$ or $1\frac{1}{2}$ or $2\frac{1}{4}$.

The significance of an average or ratio, then, depends upon three factors: the size of the population or sample; whether the statistics are discrete or continuous; and the range of the statistics over the population or sample, which is related to the second factor.

Internal and external weights

Of these, the first factor is the most important, and it is this that weighting takes care of: the greater the size the greater the weight. Suppose it is desired to calculate the average number of cars per household in a number of villages and towns in a certain locality at a particular time. We might have something like this:

Village or town	*No. of households*	*Average No. of cars per households*	*(1)* × *(2)*
	(1)	(2)	(3)
A	200	0·40	80
B	500	0·50	250
C	20	0·10	2
D	300	0·20	60
	1,020	1·20	392

The unweighted average is:

$$\frac{1{\cdot}20}{4}=0{\cdot}30$$

and the weighted average:

$$\frac{392}{1{,}020}=0{\cdot}38$$

Which is correct? Unquestionably, the weighted average is the more nearly correct. It will be seen that in order to calculate the weighted average, the procedure is: first, to adopt a system of weights based on the relative importance of the several items; secondly, to multiply each item by its appropriate weight; thirdly, to total the weights used and the products of the weights and items; and finally, to divide the sum of the weights into the sum of the products.

In the example, the system of weights consists of the number of households in each community. When the average number of cars per household is multipled by its weight, all that happens is that some of the original data, the number of cars in each community, are reproduced. The total number of cars in the four communities is 392, which is the obvious numerator to use when one is calculating the average number of cars per household in the four communities taken together. That is the essence of weighted averages: unless some arbitrary system of weighting has to be used, the products of the items and the weights are meaningful; they do not merely form a statistical stepping stone like the totals of a number of years' stock figures necessary for deriving the annual average.

Now suppose another census of population and cars owned is taken ten years later in the same four communities. It is found that each records an increase in cars owned and in the average number of cars per household; so that one would expect to see an increase in the average number of cars for all four together. In the following table the column heads have the same definitions as those in the table above.

The position 12 months later

	(1)	(2)	(3)
A	300	0·60	180
B	500	0·60	300
C	1,000	0·20	200
D	400	0·30	120
	2,200		800

Weighted average: 0·36

Although the averages for the individual communities all record increases, the weighted average shows a fall from 0·38 cars per household to 0·36. Yet there can be no doubt that both 0·38 and 0·36 are correct.

The reason for the inconsistency lies in the changed pattern of weights and the fact that the population of the community with the lowest recorded increase in the average cars per household—community C—has grown out of all proportion. It is worth noting that the unweighted average records an increase (from 0·30 to 0·43). Strictly, there is no such thing as an unweighted average. An unweighted average so called may be defined as an average for which all items are given equal weight. Where unweighted averages are used, there can be no change in the pattern of weights, and this explains why they do not produce a paradox as can happen with weighted averages where the pattern of weights is allowed to change.

The weights used in the example of the four communities are *internal* at both dates. To avoid changing the pattern of weights, that for at least one of the periods must be *external* to it, if we wish to compare the two periods' averages. The pattern for one date could be used for both, or a pattern based on the average weight (i.e. population) for each community at the two dates could be used. The following table shows the weights of the earlier period applied to the averages of the later period.

	(1)	(2)	(3)
A	200	0·60	120
B	500	0·60	300
C	20	0·20	4
D	300	0·30	90
	1,020		514

Weighted average 0·50.

The weighted average now shows an increase from 0·38 to 0·50, which is consistent with the detail. It is worth mentioning that column (3) may be defined as the total number of cars that would have been owned by the four communities had the populations remained unchanged over the ten years' interval, at the actual average number of cars owned per household at the later date.

Whether this device of using the same weighting pattern for both dates is valid depends on the purpose of the exercise. If we are concerned with the change in the average cars per household in the four communities considered as a unit, then the correct figure is a decrease from 0·38 to 0·36. If we are concerned with the change in the average of the averages, then we must either accept the increase of 45 per cent from 0·30 to 0·43 in the unweighted average, or the increase of 39 per cent from 0·36 to 0·50 in the weighted average.

It should be kept in mind that this is only an example, and that therefore no general conclusions can be drawn about the relative levels of the weighted and unweighted averages. That the former happens to be higher than the latter in the first and third of the above tables is a result of the positive correlation of the weights and averages. It happens to be lower in the second table, in which the correlation is negative. This phenomenon is exploited in Chapter 13, in order to detect whether there is a trend in any time series, and to calculate a coefficient of trend.

Weighted averages have their greatest use in making index numbers, which forms the subject of Chapter 12, where the algebra of weighted averages and of index numbers and their use in business will be found. The most common approach to the subject of index numbers is that of the shopping basket. In a week in September 1969, a housewife bought 2 lb of butter at £0·20 a lb and 3 lb of margarine at £0·10 a lb. By September 1970, the price of butter had risen to £0·25 a lb, whereas the price of margarine remained unchanged. As a result, the housewife cut down her purchase of butter by 1 lb, and made it good by an additional purchase of margarine, a process which economists call *substitution*.

		September 1969			*September 1970*	
	lbs	*price* £	*expenditure* £	*lbs*	*price* £	*expenditure* £
Butter	2	0·20	0·40	1	0·25	0·25
Margarine	3	0·10	0·30	4	0·10	0·40
Total	5		0·70	5		0·65
Unweighted average price		0·15			0·175	
Weighted average price		0·14			0·13	

The weighted averages are the averages per lb, i.e., 0·70÷5 = 0·14 and 0·65÷5 = 0·13. Again the unweighted average reflects the change in price more accurately than the weighted average; and again, the explanation is that the pattern of weights for the unweighted averages is the same in both weeks, whereas that for the weighted averages is different. Unweighted averages of prices cannot be used in making index numbers for the reason that the expenditure on some items may far exceed that on other items, and the prices of the former may be based on relatively small units of quantity. A true price index number is independent of the units of quantity and therefore of the price per unit. Potatoes, for instance, are sold by the pound and tea by the quarter of a pound. If the price of potatoes rises from £0·04 (4p) a pound to £0·05 (5p) and tea from £0·07 (7p) a quarter to £0·08 (8p), then the unweighted average price rises from $5\frac{1}{2}$ to $6\frac{1}{2}$p, an increase of 18·0 per cent. Suppose the price of tea is expressed in pence a pound, then the price would rise from 28p to 32p, and the average price of potatoes and tea would rise from 32p to 37p, an increase of 15·6 per cent.

It might be argued that if the price of every item is expressed as per pound weight, an unweighted average would be good enough. And whilst it may be possible to convert pints of milk to pounds, it would scarcely be feasible of housing. Rent, mortgage repayments and rates, which enter into retail price index numbers, are time-costs.

As demonstrated above, the main objection to weighted averages required for purposes of comparison is that the pattern of weights tends to be different in each period. Only a system of weighting in which the pattern remains constant is satisfactory. In making price index numbers, the practice, which is now almost universal, is to adopt the pattern of weights in the base period. But the weights used are not the quantities, but the money values. First, the quantities in the base period are priced at the prices in the given period and their sum is divided by the sum of the values in the base period. In the butter–margarine example above, the base-period (September 1969) quantities revalued at the given period (September 1970) prices would give a total value of £0·80:

Butter 2 lb @ £0·25 =	£0·50
Margarine 3 lb @ £0·10 =	£0·30
Total	£0·80

This, divided by the sum of the values in the base period, i.e., £0·70, gives a ratio of 1·143 or a percentage index number of 114·3.

It will be seen that the results achieved are entirely independent of the unit of quantity; for eight quarters of tea at £0·07 is precisely the same as 2 lb at £0·28. Values can be added, too, without stretching logic and common sense: one cannot add a week (for rent) to 14 pints of milk; but one can add a week's rent at £4 a week to 14 pints of milk at £0·05 a pint.

Moving averages

Whether one should use moving averages rather than moving totals is often a matter of purpose. An advantage of the latter is that it saves some arithmetic; a disadvantage, that one cannot easily compare individual items with a moving total. Where comparison forms an object of the exercise, moving averages have to be calculated. Both are used in the analysis of time series, the movement of the average or total being over time. Table 3.1 shows how they are compiled, and the additional arithmetic needed for determining the moving averages.

TABLE 3.1 Three-month moving totals and averages of sales (£'000)

		In month	*Moving total*	*Moving average*
1969	October	510		
	November	250		
	December	360	1,120	373
1970	January	240	850	283
	February	220	820	273
	March	430	890	297
	April	370	1,020	340

It was the practice some years ago to regard the moving average as a kind of norm applicable to the middle period. The average of 373, for instance, would be placed opposite November 1969, and 283 opposite December. To meet the needs of this device, the periods used were nearly always an odd number, usually 3, 5 or 7, so that there would be a middle period. There is little to recommend the notion that the moving average provides a norm by which to judge the middle period, or any other period for that matter.

Now the practice is not to regard the moving average as a norm of any particular period, but merely to take the series of moving averages as an indication of trend. In the averages of monthly figures, as in the example above, a new average becomes available every month, which gives them an advantage over figures compiled quarterly.

Like most other statistical devices, moving totals and averages need to be used with care and discretion. A recent tendency has been to use 12-month moving totals or averages to eliminate the effect of seasonal variations. It is not at all a good idea: it may eliminate the effect of seasonal variations, it also smooths out the kind of change that one might be looking for. It may hide wholly or partly a rise in costs or a fall in revenue, which on the evidence of the raw basic data could wrongly be attributed to seasonal factors. Statisticians have a metaphor for this kind of scouring process: they call it 'throwing out the baby with the bath water'. Seasonal variations are not to be removed in this way. A method of doing so is described and demonstrated elsewhere.[1]

THE GEOMETRIC MEAN

As a statistical type, the GM has tended to fall out of use in recent years. The AM of 6 and 0 is 3, the GM is 0, and this may explain why it is now less popular. Wherever one of the items in a series to be averaged is zero, the GM is necessarily zero, since any number multiplied by nought is nought, and the *n*th root of nought is likewise nought. The GM of two numbers is the square root of their product; of three numbers, it is the cube root; and of *n* numbers it is the *n*th root of their product. Except where the numbers are alike, it is invariably smaller than the AM, the extent of the difference depending upon the degree to which the numbers vary. If they are similar or approximate closely to each other, the difference is less than where they vary appreciably. The AM of 50 and 50 is 50, so also is the GM; the AM of 98 and 2 is likewise 50, but the GM is only 14.

The geometric mean has a pseudo-scientific ring about it. In most contexts, it is meaningless, and it rarely holds any advantage over the arithmetic mean. All the index numbers I know of that were ever based on the GM have either been abandoned or converted to the AM. The General Strike of 1926 made nonsense of at least one index number based on the GM. It was the *Economist* newspaper's index of business activity. During the strike one or more of the business activities covered fell to zero, so that the combined index as calculated likewise fell to zero, indicating that there was no business activity at all in the period, which was as absurd as it was untrue. There was a choice of two ways, both arbitrary, out of the difficulty: one was to count zero as one, and

[1] E. J. Broster, *Planning Profit Strategies*, Longman, 1971, Chapter 4.

the other to omit the zero items from the index for the time being. It seems the *Economist* chose the former. The statistician would argue, 'You either apply the GM to your index, or you use some other criterion. Whatever you do, you must abide by the rules appropriate to your criterion. It is the criterion that matters. The GM says in effect that your index of business activity fell to zero in the period. According to your GM criterion that is the only perfectly valid answer.' But we ordinary people must be permitted to use our common sense. We know that business activity in the period did not come to a complete halt. That it did according to the GM criterion, suggests that the criterion is unsound for the purpose.

However, the GM has its uses. The census of population is taken once every ten years. Populations tend to grow in geometric progression, so that if one wishes to make an estimate of the population midway between two censuses, the GM is the average to use. Suppose a town has a population of 20,000 in 1951, and of 40,000 in 1961, the GM of these two figures provides an estimate of the population in 1956:

$$G=\sqrt{40\times 20} \quad \text{in thousand}$$

$$=28{,}300$$

Any geometric progression conforms to the basic formula of compound interest, which every schoolboy knows. The formula is:

$$A=P\left(1+\frac{R}{100}\right)^n$$

where A is the amount at the end of n years for a principal, P, invested at R per cent compound. Let $r=R/100$, then:

$$A=P(1+r)^n \qquad \ldots \text{(i)}$$

which is the equation commonly used by actuaries and others. If the rate of interest is 8 per cent, then $r=0{\cdot}08$ and $1+r=1{\cdot}08$.

In order to determine the value of any term in a geometric progression, two terms of which we know, we need first to calculate the value of $1+r$, which is called the common ratio. From (i), we have:

$$(1+r)^n=A/P$$

$$\therefore 1+r=A/P^{1/n} \qquad \ldots \text{(ii)}$$

and

$$\log(1+r)=\frac{1}{n}\log(A/P) \qquad \ldots \text{(iii)}$$

For the town with a population of 20,000 in 1951 and 40,000 in 1961, we therefore have:

$$\log (1=r)=\frac{1}{10}\log\frac{40}{20}$$

$$=0{\cdot}0301$$

$$\therefore 1+r=1{\cdot}072$$

To determine the value of the term midway between 1951 and 1961, that is, the year 1956, we apply the formula of equation (i). Using G to represent the estimate for 1956, we have:

$$G=P(1+r)^{n/2} \qquad \text{. . . (iv)}$$

$$=20(1{\cdot}072)^5 \text{ in thousands}$$

$$=28{,}300$$

which is equal to the GM of 20,000 and 40,000 as calculated above.[1]

It will be seen, then, that even for this purpose, the GM is of limited value. It is an easy method of calculating the midway term between two known terms of a geometric progression where there is a midway term, but unlike the formula of equation (i) above, it cannot be used for calculating or estimating the value of any other term intermediate or outside the known range. An estimate for 1963, for instance, of the population of our hypothetical town can be made by applying equation (i):

$$20(1{\cdot}072)^{12} \text{ in thousands}$$

$$\text{or} \quad 40(1{\cdot}072)^2 \text{ in thousands}$$

both of which give 45,950. But it cannot be done by applying the GM. The practical value of equations (i), (ii) and (iii) in forecasting will be discussed in Chapter 13. See equations (i) to (iii) of Chapter 7 for another approach to equations (i) to (iii) above.

[1] That the GM of A and P, representing two terms in a geometric progression, is equal to the value of the term midway between A and P can be proved algebraically:

$$G=\sqrt{AP}$$

For A substitute its value in equation (i) of the text:

$$G=(P^2(1+r)^n)^{1/2}$$

$$=P(1+r)^{n/2}$$

which is the same as equation (iv).

THE HARMONIC MEAN

If a car records its first 2,000 miles at 40 miles to the gallon of petrol and its second 2,000 at 30 miles to the gallon, the average mileage to the gallon is not 35. We can calculate the correct average by extension:

$$\text{Petrol consumption in first 2,000 miles} = \frac{2{,}000}{40} = 50$$

$$\text{Petrol consumption in second 2,000 miles} = \frac{2{,}000}{30} = 66{\cdot}67$$

Total: 116·67

$$\text{Average m.p.g. is therefore } \frac{4{,}000}{116{\cdot}67} = 34{\cdot}3$$

In effect, this is the harmonic mean of 40 and 30, and it can be calculated directly without first calculating the total petrol consumption:

$$\frac{2}{\frac{1}{40}+\frac{1}{30}} = \frac{2}{\frac{3+4}{120}} = \frac{2\times 120}{7} = 34{\cdot}3$$

The general formula for the harmonic mean is:

$$\text{HM} = \frac{n}{\Sigma\left(\frac{1}{x}\right)}$$

where n is the number of terms to be averaged; Σ, the summation sign, and x, the values to be averaged.

Care is needed in using the HM. For instance, if the car covered the first 4,000 miles at 40 m.p.g. and the next 2,000 miles at 30 m.p.g., this counts as three values for averaging by the HM:

2,000 at 40 m.p.g.
2,000 at 40 m.p.g.
2,000 at 30 m.p.g.

$$\frac{3}{\frac{1}{40}+\frac{1}{40}+\frac{1}{30}}$$

$$= \frac{3}{\frac{10}{120}} = \frac{120\times 3}{10} = 36 \text{ m.p.g.}$$

This can be checked by extension.

	galls.
4,000 miles at 40 m.p.g. took	100
2,000 miles at 30 m.p.g. took	66·67
Total	166·67

$$\text{Average: } \frac{6{,}000}{166{\cdot}67} = 36 \text{ m.p.g.}$$

The weighted harmonic mean

In effect, the HM introduces a form of weighting into the formula. For this latter example, we have:

$$\frac{3}{\frac{1}{40}+\frac{1}{40}+\frac{1}{30}} = \frac{3}{\frac{2}{40}+\frac{1}{30}}$$

which brings out the weighting system more clearly.

The HM is sometimes used for averaging speeds. Suppose an aeroplane flies from A to B, a distance of 100 miles at a speed of 200 m.p.h.; from B to C, a distance of 300 miles at 250 m.p.h., and from C to D, a distance of 200 miles at 160 m.p.h., then we have:

Time taken on each leg:

	hours
100 miles at 200 m.p.h.	0·50
300 miles at 250 m.p.h.	1·20
200 miles at 160 m.p.h.	1·25
600	2·95

$$\text{Average speed for whole journey} = \frac{600}{2{\cdot}95} = 203{\cdot}4.$$

Counting 100 miles of journey as a unit of weight, we have the HM as follows:

$$\frac{6}{\frac{1}{200}+\frac{3}{250}+\frac{2}{160}} = \frac{60{,}000}{295} = 203{\cdot}4$$

One practical use which the HM is sometimes put to is in deriving Paasche's index number. The aggregative form of Paasche for a price index number, P, is:

$$P_{no} = \frac{\Sigma(p_n q_n)}{\Sigma(p_o q_n)}$$

where p is the price; q, the quantity; and the subscripts o and n refer to the base and given periods respectively. In practice, quantity data are more difficult to come by than price and value data. $\Sigma\,(p_n\, q_n)$ is the total value of purchases or sales in the given period; but to derive $\Sigma(p_o\, q_n)$, it would be necessary to have quantity data for the given period. The use of the harmonic mean provides the solution:

$$P_{no} = \frac{\Sigma(p_n q_n)}{\Sigma\left(\frac{p_o}{p_n}\, p_n q_n\right)}$$

It will be seen that p_n in the denominator cancels out, leaving $p_o\, q_n$. The term p_o/p_n is the reciprocal of the price relative, p_n/p_o. For a quantity index, Q, the price relatives themselves are used, not their reciprocals, and the value in the denominator is that of the base period not the given period:

$$Q_{no} = \frac{\Sigma(p_n q_n)}{\Sigma\left(\frac{p_n}{p_o}\, p_o q_o\right)}$$

Although, as we shall see in Chapter 12, Paasche's index is as good as Laspeyres's, it is Laspeyres's formula that has been adopted almost universally for making national retail price and wholesale price index numbers.

THE MEDIAN

The median is the middle term of a series arranged in order of magnitude. Strictly, a median cannot exist where there is an even number of terms. However, one of the two middle terms or the average of the two is usually adopted as the median. In some contexts in which I have seen it used, it is the lazy man's average. It is particularly useful where each term consists of a set of observations; for instance, the median firm in an industry decided by reference to some criterion of size such as turnover, number employed, or net assets, and setting out details of the other criteria of size and additional data on such things as the number of product lines, annual costs of labour, materials, distribution, loan interest, etc., creditors, debtors, bad debts, and so on and so on.

The adoption of the median firm as representative or typical of the industry may mislead to some extent, but it could save a great deal of time on arithmetic.

The term that lies midway between the smallest item and the median is called the lower quartile, and the one that lies midway between the median and the largest item in the series is called the upper quartile. The whole concept of the quartile, decile and percentile, which scarcely need defining, is academic. Even the median itself is of little practical use in management statistics. That its value falls between the arithmetic mean and the mode where there is any difference, is an interesting item of knowledge rather than a valuable one.

THE MODE

In any series of statistics other than time series with a definite and continuous trend, there is a bunching tendency. The item occurring most frequently, the one at the centre of the hump of the distribution curve, is the mode. For some purposes it can be regarded as more representative of a statistical series than the arithmetic mean. In a normal distribution the mode, the AM and the median are the same. In a left-hand skew, the mode is smaller than the AM; in a right-hand skew it is larger; and the median falls between the two in both types of skewed distribution.

Some but very few distributions have two humps, and are said to be bi-modal. It was discovered some years ago that the distribution of the values per ton of the various types and forms of iron and steel exported from the United Kingdom was bi-modal. But nobody ever discovered why or what significance could be attached to it, and it was dismissed as irrelevant. It is safe to say that where a distribution turns out to be bi-modal, one hump will always reach a higher level than the other. Whether the higher one or the AM of the two should be adopted as the true mode where a single representative figure of the modal type is required depends upon the circumstances, and the purpose of or the use to which it is intended to put the figure.

It is worth mentioning here that statisticians usually regard the omission from an analysis of items, readings or other entities that lie out in the blue as indefensible. They argue that every single entity should be given an equal chance to every other entity of influencing the results of an analysis. That there must be exceptions to a rule like this goes without saying: an entity may be out in the blue for some known special reason which happens to be irrelevant to the investigation, when it may be advisable to omit it.

However, the acceptance of the mode or the median as representative means abandoning all other entities, even though most of them lie within a reasonably acceptable range. If it is difficult to justify the omission of entities that lie out in the blue, how much more is it to justify the omission of most of the entities that lie within a reasonable range as well? We may be able in any event to dismiss the median as neither fish, fowl nor good red herring; but what of the mode? Unquestionably, the mode has some useful applications; even so, the wise analyst always keeps in mind the equal-chance rule when he is considering the mode as a potential indicator.

DISPERSION

A measure of the representativeness of an average is provided by the extent to which the actual items in the series deviate from it. There are two recognised measures of dispersion and therefore of representativeness. One is called the *average deviation* or the *mean deviation* which is the AM of the deviations with signs ignored, and the other the *standard deviation*, which is just as much an average as the former, albeit more sophisticated. The standard deviation is said to conform to the theory of error, and is more commonly used than the average deviation.

Table 3.2 contains a short series of figures, enough for the purpose, showing how the two kinds of deviation are calculated. There is not much difference between the two measures of dispersion, the average deviation being 14, and the standard deviation 17·3. It is worth mentioning that the standard deviation always exceeds the average deviation.

Greek letters are generally used to represent derived statistics. The one most commonly used for the standard deviation is small sigma, the formula being:

$$\sigma = \sqrt{\frac{\sum x^2}{n}}$$

where x represents the deviation from the mean of a series of n items.

Deriving the deviations from the mean of a long series can be tiresome. In calculating the standard deviation, it can be avoided by taking the squares of the individual values of X as shown in column (1), correcting the sum of these squares by deducting the product of the sum of the X series and the average of X, to give the sum of the squares of the deviations. Column (4) of Table 3.2 gives the squares of the X series; the

TABLE 3.2 Deriving the average and the standard deviations

	Basic series X	*Deviations from mean* x *(1) − 36*	*Deviations squared* x^2	*Basic series squared* X^2
	(1)	(2)	(3)	(4)
	10	−26	676	100
	15	−21	441	225
	42	6	36	1,764
	53	17	289	2,809
	70	34	1,156	4,900
	41	5	25	1,681
	29	−7	49	841
	36	0	0	1,296
	44	8	64	1,936
	20	−16	256	400
TOTAL	360	140*	2,992	15,952
AM	36	14	299·2	
Square root			17·3	

* Ignoring signs.

procedure is demonstrated as follows:

$$\sum(X^2) = 15{,}952$$

$$\sum X \times \frac{\sum X}{n} = 360 \times 36 = 12{,}960$$

$$\text{Difference} = \sum(x^2) = 2{,}992$$

which is the same as the value of $\sum(x^2)$ shown at the foot of column (3). The expression $\sum X \times \frac{\sum X}{n}$ may be written $\frac{(\sum X)^2}{n}$. The difference between $(\sum X)^2$ and $\sum (X^2)$or $\sum X^2$ is worth noting: in the former it is the sum of the X's that is squared; in the latter it is the individual values of X that are squared.

Another name for the standard deviation is the *root-mean-square deviation*, which is a term used by electrical engineers;[1] and the square of the standard deviation, i.e. 299·2 in the example of Table 3.2, is called the variance.

[1] Sometimes written RMS or $\sqrt{MS}$.

If the values of X are expressed in tons, then the AM, the average deviation and the standard deviation are also expressed in tons; but the variance is expressed in tons squared, whatever that may mean. A form of the standard deviation independent of the unit of value or quantity is the coefficient of variation: it is the ratio of the standard deviation to the AM; or the percentage coefficient of variation: the percentage that the standard deviation bears to the AM. For the example of Table 3.2, the coefficient of variation is:

$$\frac{17{\cdot}3}{36} = 0{\cdot}481 \text{ or } 48{\cdot}1 \text{ per cent}$$

Whereas a standard deviation has no meaning outside the context of the series it relates to, a coefficient of variation has a significance within itself and can be compared with any other coefficient of variation. One can appreciate the dispersion of a series without knowing the basic series itself, or the terms in which it is expressed.

There are some extremely simple measures of dispersion all consisting of ranges, viz.:

1. *The full range*, from the lowest value to the highest in the basic series;
2. *The interquartile range*, from the lower quartile value to the upper quartile value;
3. *The interdecile range*, measured from the first decile to the last, thus omitting the first and last tenths of the series;
4. *The interpercentile range*, measured from the first percentile to the last, thus omitting the first and last hundredths of the series.

These may be useful for purposes of presentation, but like the average deviation, they are of little if any value for analysis.

The advantage of the standard deviation is that under normal conditions, it provides the analyst with a measure of the probability that any observation will fall within it or any multiple of it, and therefore of the probability the observation will fall outside it or a multiple of it. There is a 68 per cent probability that an observation will fall within one standard deviation of the AM, and therefore a 32 per cent probability that it will fall outside it. For half a standard deviation, the probability of inclusion is 38 per cent; for $1\frac{1}{2}$, it is 87 per cent; for 2, it is 95 per cent; and for 3, 100 per cent.

Needless to say, these figures are based on large numbers of items. For the example of Table 3.2, which is very small, the actual results are not outrageously dissimilar. They are:

No. of standard deviations	*Range covered*	*No. of items included*	*Proportion of total* %	*True probability* %
$\frac{1}{2}$	27–45	5	50	38
1	19–53	7	70	68
$1\frac{1}{2}$	10–62	9	90	87
2	1–71	10	100	95

For determining the range within which an additional item would fall for a given probability, the figures in the last column should be used. If the example applies to the tonnages of miscellaneous goods traffic despatched from a particular goods depot on ten consecutive working days, there is a 68 per cent probability that the tonnage presented for despatch on the eleventh day will fall within one standard deviation of the average, that is, in the range 19–53 tons. It is worth keeping in mind that a 50 per cent probability corresponds to three-quarters of the standard deviation.

A NOTE ON SQUARING

As the squares of numbers play an important part in statistical analysis, a note on short-cut methods of calculation may be useful here. An electronic computer or desk calculator is not always available for the odd calculation, or convenient to use even though it be available.

If we have the square of a number and we need the square of the next number higher all that is necessary is to add twice the lower number plus one to the square. Similarly, if we need the next number below, we deduct twice the higher number from the square and add one. A simple example will make this clear.

$$12^2-11^2=144-121=23$$

$$11\times2+1=23$$

$$12\times2-1=23$$

Let X be the lower number; then $X+1$ is the number next above it; and

$$(X+1)^2=X^2+2X+1$$

The idea can be extended to numbers that are not consecutive. Let d be the difference, then:

$$(X+d)^2=X^2+2dX+d^2$$

If the first number in a series is 240, and the second is 245, we can calculate the square of 240 by starting with a simple square, that of 200, and applying the formula of the equation:

$$(200+40)^2=40{,}000+2\times 40\times 200+1{,}600$$
$$=57{,}600$$

Then we proceed to calculate the square of 245:

$$(240+5)^2=57{,}600+2\times 5\times 240+25$$
$$=60{,}025$$

If the second figure in the series had been 235, then the appropriate formula is:

$$(X-d)^2=X^2-2dX+d^2$$
$$(240-5)=57{,}600-(2\times 5\times 240)+25$$
$$=55{,}225$$

For a series of numbers, where a calculator or a table of square is not available, a tabular format can be used, e.g.:

X (1)	d (2)	$2dX$ (3)	d^2 (4)	(3)+(4) (5)	X^2 (6)	$(X+d)^2$ (5)+(6) (7)
200	40	16,000	1,600	17,600	40,000	57,600
240	5	2,400	25	2,425	57,600	60,025
245	1	490	1	491	60,025	60,516
246					60,516	

The squares of $(X+d)$ are carried back and down to column (6) in the line below. One important advantage of this method is that a single check, that of the final square, provides a check on the whole table. Thus, in the example, the square of 246 can be calculated independently. If it proves to be 60,516 as shown in column (6), it can be taken that the arithmetic of the whole table is free of error. There is no scope for compensating errors in calculations of this kind. One further point: in tables showing successive differences like those in column (2), it is usual to show them as applying to the second number of the pairs of successive numbers. However, in this exercise, it is convenient to show them against the first number of the successive pairs. In practice, the table is a working document, not one for presentation.

There is another way of reducing the amount of arithmetic. Many statistical series consist of numbers running into five or six significant figures, yet the range may be quite narrow, perhaps not more than three significant figures. Rounding can safely be carried out where the range is relatively broad, but not where it is narrow. The method is to deduct from each figure the lowest in the series and treat the new series thus derived as appropriate for the squaring process. A simple correction completes the work of deriving the sum of the squares of the deviations from the mean. Table 3.3 contains a worked example. The correction is the square of the sum of the deviations from the lowest figure, divided by thc number of observations. In the example, it is $773^2/8 = 74{,}691$, which is deducted from the sum of the squares of the deviations from the lowest to give the sum of the squares of the deviations from the mean of the original series as shown in column (3). The slight difference between the two sums of squares as calculated is due to rounding the average of the original series in column (1), and the consequential

TABLE 3.3 Deriving the sum of the squares of the deviations from the mean

	Original series X (1)	*Deviations from mean* $X-1{,}029{\cdot}6$ (2)	*Squares of deviations from mean* (3)	*Deviations from lowest* $X-933$ (4)	*Squares of deviations from lowest* (5)
	1,068	38·4	1,474·56	135	18,225
	1,055	25·4	645·16	122	14,884
	1,118	88·4	7,814·16	185	34,225
	1,032	2·4	5·76	99	9,801
	937	−92·6	8,574·76	4	16
	933	−96·6	9,331·56	0	0
	986	−43·6	1,900·96	53	2,809
	1,108	78·4	6,146·56	175	30,625
TOTAL	8,237		35,893·48	773	110,585
Average	1,029·6				
Correction to col. (5): deduct $772^2 \div 8$ =					74,691
					35,894

rounding of the figures in column (3). The method demonstrated in columns (4) and (5) is therefore not only a short cut, it is also more accurate, which can be important, since in some kinds of statistical analysis the results may be highly sensitive to discrepancies both large and small.

Where the range of a series is relatively wide, there would be little if any saving in arithmetic in using the deviations from the lowest figure except possibly for use in regression analysis (see Chapter 7). To avoid the inaccuracies frequently involved in using the deviations from the mean, the method demonstrated in column (4) of Table 3.2 can be used, i.e., the figures of the basic series themselves are squared, summed and corrected. For the example of Table 3.3, the sum of X^2 is 8,516,915 and the correction is 8,237 squared and divided by 8 = 8,481,021, which, deducted from $\sum(X^2)$, gives 35,894, the same as the figure at the foot of column (5).

4

Sampling

Sampling is fundamental in statistical science. An average can be regarded as representative of a sample of a universe, and its standard deviation as a measure of the extent to which any observations drawn from the universe inside or outside the sample will approach the average.

There are good samples and bad samples. A good sample is one that is representative of the universe; a bad sample, one that is not. Statistical theorists often seem to suppose that obtaining a representative sample is a matter of size and care in selection. In practice, samples are all too often very small and self-selected. In the context of forecasting, the universe, or *population* as it is often called, lies partly in the future. But the sample available necessarily lies in the past and present: not a single item can be selected from the future. An oft-quoted rule of sampling is that every item in the universe must have an equal chance of selection. Items lying in the future have no chance at all.

However, not all universes subject to sampling have a part in the future. And for those that have, there is no alternative to ignoring the future part of them.

Before embarking on a sampling exercise, the investigator needs to find the answers to a number of questions and to make a number of decisions about the sampling answers being sought, the nature of the universe to be sampled, how to go about selecting the sample, the size of sample, and so on. These are discussed in the following sections.

THE SAMPLING ANSWERS

What precisely do we want? What is the purpose of the exercise? Do we want to know the proportion of households that have refrigerators? Or do we want also a classification by reference to type and size of refrigerator? Do we want to know the proportion of electors who favour the Government? Or do we also want a classification of electors by reference to sex or to type of constituency? Do we want the value of net assets or the investment expenditure of public companies? Or do we need an analysis by reference to size of company? What should be the criterion of size? The purpose of the sample enquiry should be clear in the mind of the investigator before he begins his work. That the purpose should not be to prove a theory goes without saying, as it does with all statistical enquiries and indeed with scientific enquiry of any kind. A preconceived theory all too often tends to colour the results. Open-mindedness is an important attribute of all good investigators.

Admittedly a hypothesis is often a necessary part of purpose in scientific investigation; but it should be expressed in general terms, and avoided as far as possible where the object is to prove or disprove a theory. Many statistical investigations are carried out to obtain a measure of, say, the effect that one factor has upon another. There is necessarily a hypothesis, but the underlying theory is not one to be proved or disproved. The director of a public opinion poll may set out with the preconceived theory that the majority of the electorate support the Government; and he may not find it difficult to prove his theory. On the other hand, he may approach his task in an open-minded scientific way with the object of *measuring* the support the Government has in the country.

BUILDING THE UNIVERSE

A good universe is not always easy to come by. In some cases, the universe may exist as little more than a theoretical concept. We know what the universe should comprise, but there may not be any such physical entity. In others, it may be difficult to decide what the appropriate universe should be. A sample enquiry of a research project designed to determine the potential sales of children's jeans made of a new kind of material would set the research worker a number of problems related to his universe. The number of children by sex and age group of from 5 to 15 years old living in the test area may form a basis of potential demand. Children rarely buy their own clothing, so that they could scarcely be regarded as a universe for sampling purposes. It is parents, grandparents and other relatives who buy children's clothing, and it is they who decide whether the jeans are worth buying at the price.

The next step is to build up a suitable universe of children's clothing buyers. Needless to say, it would be an extremely difficult task. A universe of the parents of children of 5 to 15 in the test area would be difficult enough to build without taking account of grandparents and other relatives. In view of the equal-chance rule, would it be wise to omit them? Perhaps this obstacle could be surmounted by the questions to be put to the parents. Sources of information about parents are:

1. *The electoral roll*, which gives no indication of marital status. John and Mary Smith living at 21 Vicarage Road, may be man and wife or brother and sister. The man and wife may have no children of school age, and the brother and sister may have ten children of school age in their care, the brother being a widower. Similarly a woman recorded as alone on the register may be a widow or a deserted wife with young children. For the purpose of building a universe, the electoral roll would not be very helpful.
2. *Parents associations*, which may be more fruitful, but which present their own problems. Not all parents of school children are members of parents associations. And many may have schoolchildren of over 15, but none under.
3. *Children's welfare department, and the local education authority*, both of which, especially the latter, may possess a list of parents and guardians of schoolchildren in their areas with the ages of children recorded. It is questionable whether either would release such a list or lay it open for inspection by 'unauthorised' enquirers. But even if they would, there still remains the problem of private schools, children with their own tutor at home, and children at boarding schools outside the area.

Probably, the manufacturer or merchant who sells direct to consumers and never through retailers would have to have recourse to the electoral roll after all. For the one who sells through retailers, it would probably be wise to consult the retailers themselves—they are knowledgeable in such matters—rather than go to the expense of a sample enquiry based on a universe of unknown dimensions and uncertain definition.

DRAWING THE SAMPLE

A good sample is a carefully controlled random sample. This may sound like a contradiction in terms. How can one have a controlled sample that is also a random sample? The answer is simple: it is the randomness that is controlled rather than the sample itself; it is controlled to lend the maximum randomness to the sample.

Statisticians have gone to a great deal of trouble to devise methods of control for maximum randomness. One such method is to allot numbers serially to all the items in the universe, and then to draw the sample by reference to a table of random numbers, of which many have been published, some compiled manually by various means, others by computer.

However, selecting a sample from a universe is largely a matter of common sense and of keeping in mind the equal-chance rule. The investigator who makes an effort to give every item in his universe an equal chance of selection cannot go far wrong. If the universe is represented by a card index, and a 10 per cent sample is to be drawn, every tenth card should make up a good sample provided some means of randomly selecting the first one is used. Drawing one number from a hat containing the numbers one to ten for a 10 per cent sample or one to five for a 20 per cent sample and so on should satisfy the need for randomness. It should be kept in mind that this first selection is not unimportant: in effect it determines the whole sample. If the fifth item is drawn for the first item of a 10 per cent sample, then the second item would be the 15th; the third, the 25th; and so on.

An alternative but similar method is to draw one number from the hat for each group of ten items of the universe to make the selection of a 10 per cent sample. In this way, the selection of each sample item is rendered entirely independent of the selection of the first or any other item. Unquestionably, this method would involve more time and trouble than that in which every tenth item is drawn; but in a situation where there exists or where it is suspected that there exists a numerical cycle in the universe, it may be well worth while.

Where a 50 per cent random sample is drawn, the remainder of the universe provides a sample as good in all respects as the drawn sample: it is equally random and equally representative. It is true that the results of the two samples would differ one from the other; and apart from the size of the sample, the extent of the difference would depend upon the magnitude of the dispersion of the items of the universe.

Similarly, if a 10 per cent (i.e. one-in-ten) random sample is drawn, the remaining nine-tenths of the universe is as good as the whole universe for providing a one-in-nine random sample, and this second sample would be as good as the first in all respects although in fact it is drawn from an incomplete universe. Some statisticians exploit this phenomenon by dividing the sample they need into two or more equal parts, drawing each one separately, and comparing the results of the two.

The phenomenon also lends itself to sequential sampling, which is frequently used in the quality-control sampling of batches, particularly where testing to destruction is necessary as it is in the testing of electric lamps for length of life. Sequential sampling consists of drawing and analysing a small sample and if its results are inconclusive of drawing a second and so on until a reasonably definite inference can be drawn from the results.

STRATIFIED SAMPLING

In some types of sampling the standard equal-chance rule is abandoned in favour of maximising the representativeness of the sample for a minimum of effort. Where the items of a universe vary in relevant size, a sample of, say, 20 large items will be, or at any rate tend to be, more nearly representative of the universe than a sample of 20 small items. Since the whole object of sampling is to save work, every attempt to minimise the effort needed to achieve a given degree of representativeness is to be applauded.

Perhaps it is not quite true to say that the equal-chance rule is abandoned. A stratified sampling technique is adopted to make the best of varying sizes of items. The universe is first divided into groups or strata, each stratum covering a size range. A proportionately larger sample is selected from the stratum with the largest sized items than from any other stratum. The standard rules of random sampling apply to each stratum, as though it were a universe in its own right, which it may well be.

Where the universe has a left-hand skewed distribution, that is, where the small items outnumber the large items, the sampling fraction for the larger items should logically be much greater relative to that for the smaller items than for a universe which has a right-hand skewed distribution. This is not the only factor to be taken care of in deciding the range of sampling fractions. Another factor is the dispersion existing in the universe: the greater the dispersion, the greater the range should be.

Where there is a decided left-hand skew and an appreciable dispersion, the sampling fraction for the top stratum, i.e. the stratum containing the largest items, may be as much as unity, or, in other words, 100 per cent; and that for the bottom stratum as little as 100th or 1 per cent. The only limit to the number of intermediate strata is the desired size of sample. It seems the aim should be to have a sample for each stratum reasonably representative of the stratum. There would be no point in having so many strata that some of them contributed only one or two items to the sample.

For any given universe, there may be an optimum number of strata, and an optimum range of sampling fractions, for any given probability or degree of representativeness of the total sample; and it is likely that the problem has been discussed more than once at a high academic and theoretical level. As it is, in practice, the problem can best be solved by personal judgment, the main points worth keeping in mind being (*a*) that a predetermined minimum number of items, say, ten, to each stratum should be adhered to, and (*b*) that the object of stratifying is to reduce the volume of work involved in the sampling exercise without impairing the representativeness of the whole sample.

QUOTA SAMPLING

In some sampling exercises, the object is not so much to save effort as to save time. Public opinion polls or social surveys are usually of this kind, especially those designed to provide a forecast of a general election result. Market surveys, too, fall into the same category; they are in fact a form of unpublicised opinion poll, the same commercial undertakings conduct both social and market surveys; they are organised for speed. There is no time for a postal questionnaire, and the undertakings employ field workers to question people they encounter in the streets and in other public places. Controlled random sampling is out of the question. Each field worker has a quota of interviews to make in the course of a few hours, and he makes his own selection along the general lines laid down by the undertaking. Having done his quota, he summarizes the results and sends them at once by telephone, telegram or letter to his central office.

Social surveys done by quota sampling have formed the subject of numerous stories circulating amongst statisticians, some true enough, others probably pure invention. One typical story provides a good example of the risks inherent in quota sampling. It was told as a true story at a meeting of the Royal Statistical Society before the war. A survey of public opinion on horse-racing and betting was being carried out. A field worker who had been allocated an area of the City of London, embracing Bishopsgate and neighbourhood, found in Liverpool Street railway station all the people he needed to satisfy his quota of some 40 or 50 interviews. There was a long queue of men and women standing waiting for a train, and he decided he would question every fifth person in it and thus randomise his sample. Not only every fifth person but also those who overheard the questions asserted emphatically that they were all in favour of horse-racing and betting. The field worker reported the results of his enquiries; but following a credibility test of the survey, it was discovered that the train for which the queue at Liverpool Street Station was waiting was a racecourse special destined for Newmarket.

BIAS

A more apt prologue for a section on bias than the story of the field worker at Liverpool Street Station can scarcely be imagined. Unwittingly, the field worker had selected the most biased sample possible. Extreme bias like this is probably less harmful than a gentle bias. A credibility test consisting of a comparison of each field worker's return with the rest, and if possible with the full results of an earlier survey immediately brings to light the probability of bias where it is extreme, rarely, if ever, when it is gentle. Each of the stories told, whether true or not, teaches its own lesson to form the basis of a principle of survey sampling. Most of a code of principles and instructions thus compiled would be concerned with the avoidance, or, at any rate the minimisation, of bias.

The difficulty with bias is that it is unmeasurable. The standard sampling error formula assumes there is no bias embodied in the sample, except for a small sample, for which there is a standard formula for correcting the sampling error to allow for smallness of sample. Small samples tend to have more bias than large samples. Indeed, large samples are mainly justified on the grounds of their lack of bias.

It is often thought that bias in a sample is the result of poor sampling. There is reason for supposing that it can result from a poorly contrived universe as well. It can be said that there are two concepts of universe, one the ideal, which is often purely theoretical, and the other the available, which is practical. A survey carried out to ascertain the general opinion of the National Health Service, for instance, might have as its ideal universe of members of the public, the people on the electoral rolls. The available members of the public at any particular time would exclude people abroad, in hospital, jail and other institutions; and people who are in process of moving household, some of whom are semi-nomads and never appear on an electoral roll. The omissions could cause some bias. People who spend some time abroad have a standard of comparison, people who have spent some time in hospital may on the whole have an entirely different opinion of the N.H.S. than people who have not; both are in a better position to express an opinion than other people, though that may not be of any importance.

Probably the most serious bias arises in postal surveys. The ideal universe may be the electorate for a public opinion poll; or the members of a trade association for a survey concerned with average turnover, net assets, work-force or payroll in a particular industry. The available universe consists of the people or companies which would respond to a postal questionnaire. As statisticians have pointed out many times, the omission from the universe of non-responders introduces an unknown bias into the sample. Responders and non-responders as groups each have their peculiarities, and these peculiarities may be relevant to the investigation. The problem is that nobody knows these peculiarities. It is also a fact that responders almost invariably account for no more than 50 per cent of the sample. A great deal depends upon the sponsor. If the sponsor of the enquiry is a Government department, the response will be greater than if it is, say, a department of a university. A graduate of a university seeking data for a thesis will receive scarcely any response at all. It is not enough to double or treble the size of a sample to make good the non-responders: that would not eliminate the bias.

THE SAMPLING ERROR

Where a sampling enquiry is designed to determine an average, such as the average yield of wheat per acre, output per man-hour of a large labour force engaged on similar work, or the fuel consumption per mile of a large fleet of road motor vehicles, it is often important to determine the error in the average of the sample, to find the extent to which the sample average is likely to deviate from the average of the universe for a given probability. In effect, what it seems is needed is the standard deviation of the averages of a number of samples drawn from the universe.

But there is no need to draw more than one sample for the purpose. First, the standard deviation of the items in the sample is calculated, that is, for instance, of the yields per acre, or the outputs per man-hour observed in the sample. This is divided by the square root of the number of items in the sample less one to give the sampling error for a 68 per cent probability. In short, then, the dispersion of the items in the sample as measured by the standard deviation provides the clue to the sampling error. To save any ambiguity, the formula is given as follows:

$$\text{Sampling error} = \frac{\sigma}{\sqrt{n-1}}$$

where small sigma, σ, represents the standard deviation of the items in the sample; and n is the number of items in the sample. If the average yield of wheat per acre in a locality shown by a sample of 50 items is 40 cwt, the standard deviation being 5, then the sampling error would be:

$$\frac{5}{\sqrt{50-1}} = 5/7 = 0{\cdot}714 \text{ cwt}$$

This can be interpreted in the same way as the standard deviation as discussed in Chapter 3. It can likewise be doubled to give a 98 per cent probability that the averages of other samples drawn from the same universe and the average for the universe as a whole would fall within the range (i.e., in the example, $40 \pm 1{\cdot}428$ cwt) and so on as in Chapter 3.

It should be kept in mind that the formulae of the standard deviation and the sampling error take care of chance deviations only. They are based on the assumption that the sample is not biased. They also assume that the frequency distribution of the items and of the averages is a normal one. However, except where the sample is self-selected, but where otherwise, reasonable care is taken in circumscribing the universe and in selecting the sample, both parameters, especially the sampling error, can be of great practical value.

In a spot-check of items in, for instance, the audit of a company's books of account, neither parameter can be of any use. The object of a spot check is to test individual items for accuracy, not to determine the average or the total of a universe of items. All that is necessary is to make a random selection of items to be checked; there can be no question of bias. Stratified sampling is a useful technique for auditing purposes especially where the number of items forming the universe is large and where the dispersion is appreciable. Of course, there is no reason why an auditor should not check a total by calculating the average of a sample and the sampling error and multiplying the two by the number of items in the universe to give him a total and error margin. If there is any objection at all, it is that the exercise might be somewhat academic.

5

Micro-economics and model building

Micro-economics means literally the little economics, *micro* from the Greek *mikros* small. Macro-economics means the large economics, *macro* from the Greek *makros* long or large. Macro-economics is current jargon for the study of the wealth of nations, to use Adam Smith's apt and idiomatic phrase; and micro-economics is, by analogy, jargon for the study of the wealth of firms or industries. Both consist of two elements, theory and empiricism. This chapter is concerned with the wealth of businesses and how the theory or hypotheses of business management can be made subject to statistical analysis through the mathematical model.

There are always constraints to take care of in applying theory to practice. These constraints provide pitfalls for the unwary. They exist particularly in expansion situations, though it would not be true to say that there are no constraints in situations of contraction. A hypothetical example will demonstrate this proposition.

CONSTRAINTS

Management accounting literature has in recent years contained many demonstrations of break-even analysis, and break-even sales in particular. The demonstrations are usually accompanied by a diagram showing the graphs of the total annual cost of producing an individual commodity and the corresponding sales proceeds, for a range of rates of output and quantity sold. Costs and proceeds, £, are measured on the vertical axis and physical quantities, Q, on the horizontal axis, on the lines of Fig. 5.1.

Both graphs are represented as straight lines rising to the right, the cost graph beginning part way up the vertical axis from a point which represents the annual fixed cost, F; and the proceeds graph from the point of origin, O. The point of intersection, *B*, of the two graphs indicates the break-even sales proceeds on the vertical axis and the quantity on the horizontal axis. Usually, the area between the two graphs below and to the left of the break-even point is shaded by, say, vertical lines and is labelled LOSS, and the area above and to the right of the break-even point is shaded by, say, horizontal lines and is labelled PROFIT.

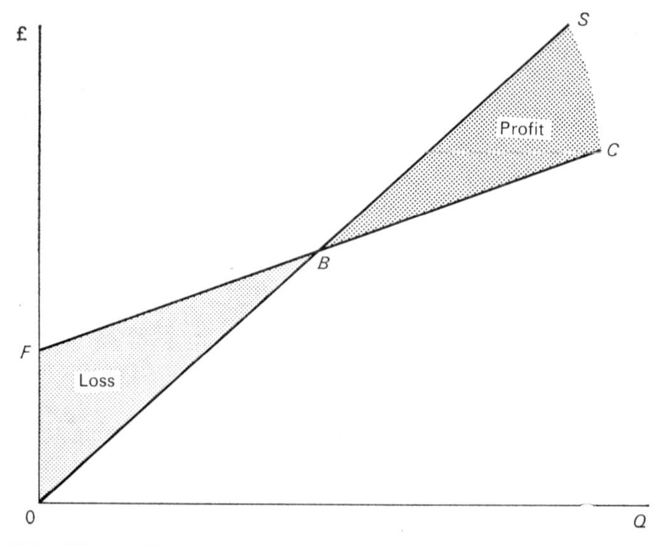

FIG. 5.1 The static break-even chart

There are two main constraints to be taken into account in this kind of analysis. One is the existence of a limit to the capacity of the plant and also to the labour and raw materials available, and the other is the existence of a limit to the size of the market. In short, neither graph can be regarded as continuing indefinitely to the right in a straight line. The cost graph may change in three ways: (*a*) it may rise bodily as a result of a rise in fixed costs caused by an addition to the capacity of the plant; (*b*) it may begin curling upward owing to the special efforts called for to increase the rate of production beyond the normal capacity of the existing plant and (*c*) it may rise to the right at a steeper slope owing to a rise in wage rates necessitated by an increase in the labour force and to a rise in the prices of materials caused by an increase in the firm's demand for them. Admittedly, this last is somewhat theoretical.

Indeed, an increase in the requirement of raw materials could result in bulk buying and a lower price.

There is only one way by which the sales proceeds graph could change, and that is for it to become less steep. This kind of analysis, as demonstrated in Fig. 5.1, assumes a fixed price. The rate of sales at a fixed price is limited by the price, and in any expansionary process, the time would come when any further expansion could be achieved only at the expense of a fall in the selling price.

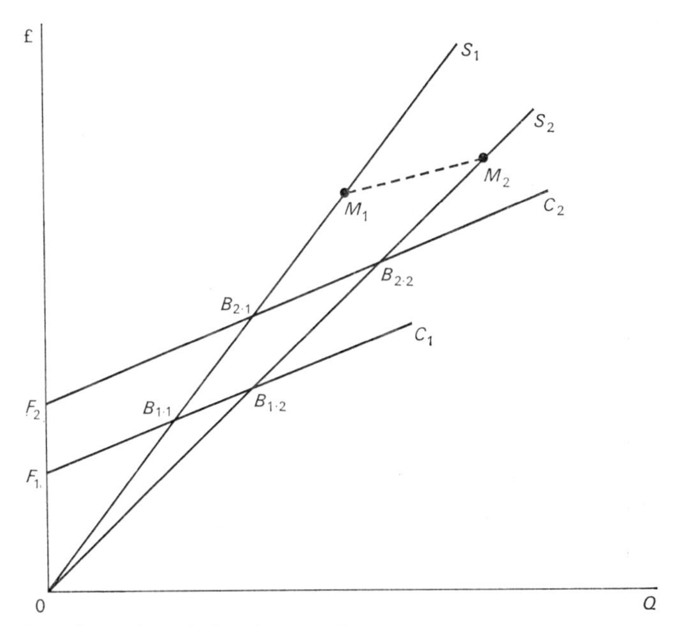

FIG. 5.2 A semi-static break-even chart

Figure 5.2 shows the original graphs of annual costs and sales as in Fig. 5.1, and their new positions and slopes following an increase in the capacity of the plant and a reduction in price. F_1 and F_2 on the vertical axis represent the annual fixed costs under the original and expanded capacities respectively. The slopes of the two cost graphs are the same on the assumption that the marginal cost remains unchanged. Graph S_1 represents the sales at the original fixed price and S_2 the sales at the lower fixed price. The slopes of the two graphs are the measure of the two prices. The points M_1 and M_2 represent the annual sales at the two prices. M_2 is shown as standing at a higher level than M_1, that is, the lower price yields greater annual sales proceeds than the higher price, a situation which is normal in practice.

THE TRUE SALES GRAPH

The true sales proceeds graph allows for changes in price. It rises steeply at first, concave downward, from the origin and it would pass through the points M_1 and M_2, on what must now be defined, not as sales graphs, but as price lines and unit cost lines. The prices and unit costs could be calibrated on a quarter-circle scale beginning at a high level on the vertical axis, and terminating on the horizontal axis. As an understanding of this is important in micro-economics and the wealth of firms, it is diagrammatically represented in Fig. 5.3, which is drawn so as to show how the price and unit cost scale widens its calibrations as the price and unit cost falls.

It will be seen from Fig. 5.2, that there are four break-even points, lettered $B_{1\cdot1}$, $B_{1\cdot2}$, $B_{2\cdot1}$ and $B_{2\cdot2}$. It should be kept in mind that these all assume that the price is fixed at each of two figures in turn, and that it

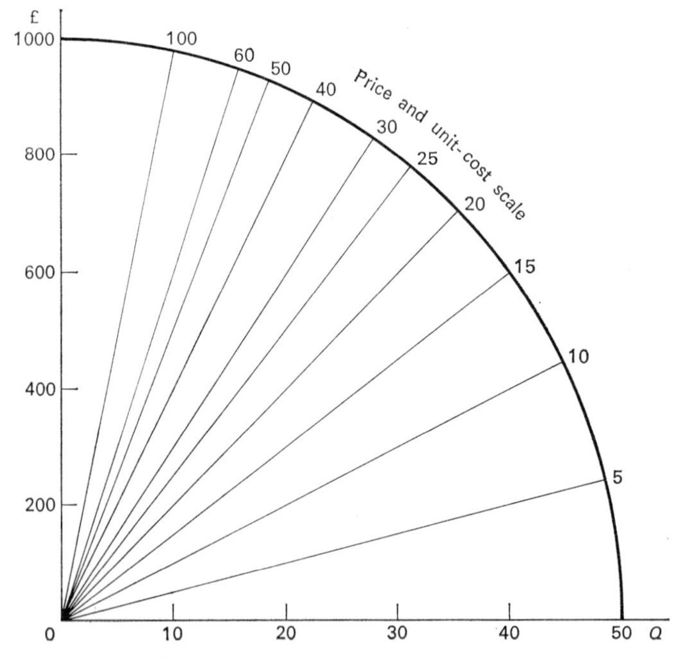

FIG. 5.3 Price and unit-cost lines radiating from the origin

is not regarded as a variable factor in the rate of sales. There are also two cost graphs represented in the diagram. The number of break-even points in a diagram of this kind is equal to the product of the number of cost graphs and the number of sales graphs or price lines. It is a less static form of the break-even chart usually presented in the literature, as in Fig. 5.1.

Dynamic break-even analysis is fundamental in micro-economics and the wealth of firms. It is fully discussed and its practical value demonstrated in Chapter 19. In the meantime, it is worth keeping in mind that the commonly expounded form of break-even analysis fails to recognise the constraints which are inherent in such a static concept.

RATIONAL PRICING

The basic theory underlying dynamic break-even analysis also underlies rational pricing. It similarly takes a dynamic form and is thus less constraint-ridden than the static analysis. It can safely be said that pricing theory is the most important concept of micro-economics and the wealth of firms.

In the previous section, it was stated that the true sales-proceeds graph rises from the origin to the right and is concave downward. It passes through the point of maximum sales for each given price, e.g., the points M_1 and M_2 on lines S_1 and S_2 in Fig. 5.2. A complete theoretical sales curve would ultimately become parallel for a short distance to the horizontal axis, and then turn downward until it reached the horizontal axis for zero price. In practice, only a short section of the curve can be known and even then, it is often little more than an approximation. The length of the section depends upon the range of prices that have been charged in recent years for the brand of product or service.

One important point about this kind of representation is that the diagram invariably applies to *rates* of sales or costs in unit time, say a year, and the graph is never intended to be taken as a time trend. The graph rises to the right as the *rate* of sales or output increases, and not with the passage of time. The literature is studded with examples of misconception in which writers, some of whom ought to know better, appear to believe that the graphs of costs and sales-proceeds represent time trends, or at any rate, changes in output that take place in time. Time trends and movements that take place with the passage of time ought to be presented in diagrams whose horizontal scales provide a measure of time. It is all too easy to slip into the mistaken belief that the lower ranges of quantities shown on the horizontal axis apply to shorter periods of time than the larger quantities. If the diagram applies to annual rates and it is known that the ouput in a year is evenly distributed over the year, and if the horizontal axis measures up to 12,000 tons, even then it would be quite wrong to suppose that the first 1,000 tons shown on the axis applies to January of the year; the next 1,000 tons, to February, and so on. The first 1,000 tons on the axis represents the annual rate just as much as 12,000 tons. The point is so important and has been misunderstood so often that it is emphasised from time to time in the following chapters.

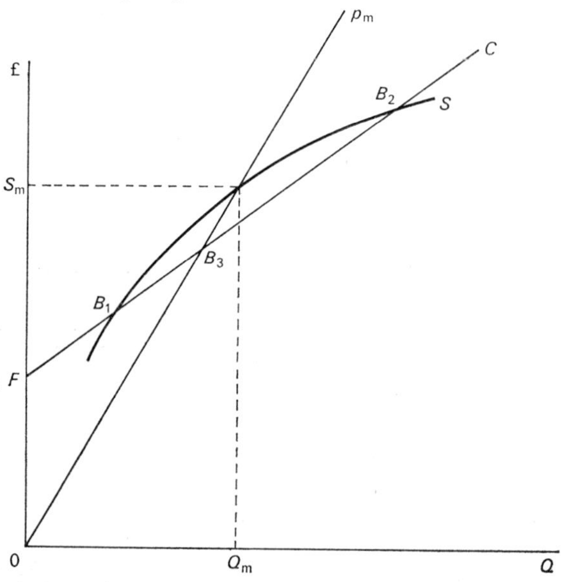

FIG. 5.4 A dynamic break-even chart with optimum price line, P_m

Figure 5.4 presents the typical graph of the annual sales-proceeds of a brand of a product as it may be found in practice. It also presents the typical graph of annual costs to complete the picture needed for rational pricing. The vertical distance for any quantity between the two curves as measured on the vertical axis represents the annual profit or loss, the former where the sales graph lies above the cost graph, and the latter where it lies below.

In theory, there are two break-even rates of sales, one for very high prices as shown in the diagram, and the other for very low prices. In practice, it is likely that neither can be determined with any degree of certainty. Sufficient for the purpose of rational pricing is the section of sales graph which runs more or less parallel to the cost graph, for it is here that the maximum distance between the two graphs and therefore the maximum profit is found, as Fig. 5.4 shows. The co-ordinate, on the sales graph of S_m and Q_m, representing the optimum annual sales and the optimum annual quantity respectively, is the point through which the optimum price line, p_m, passes, *optimum* referring to the maximisation of profits. It will be seen that a movement upward or downward of the cost graph does not affect the optimum sales, quantity or price, provided its slope remains the same. But such a movement affects the magnitude of the maximum profit. The point, B_3, where the cost graph and optimum price line intersect appears to represent a break-even situation which in this type of dynamic analysis, is of no significance. The points of intersection of the cost and sales graphs at B_1 and B_2 provide break-even situations of real significance.

MODEL BUILDING

Models are of three kinds, viz.: the verbal, the numerical and the mathematical. Verbal models fall into two classes: the chronicle and the chain of cause and effect. Both can take either of two forms: the narrative and the diagrammatic. The narrative form merely expresses the order of events chronologically or the chain of cause and effect in grammatical English; and the diagrammatic in the form of a series of boxes each representing an event, or a cause or effect, and so labelled. Figure 5.5 below is an example of the diagrammatic form of a chronicle verbal model. The diagrammatic form is often somewhat facetiously referred to as a *wiring diagram*, a term which is also applied to flow charts and similar forms.

Numerical models consist of tabulations of numerical data, or of

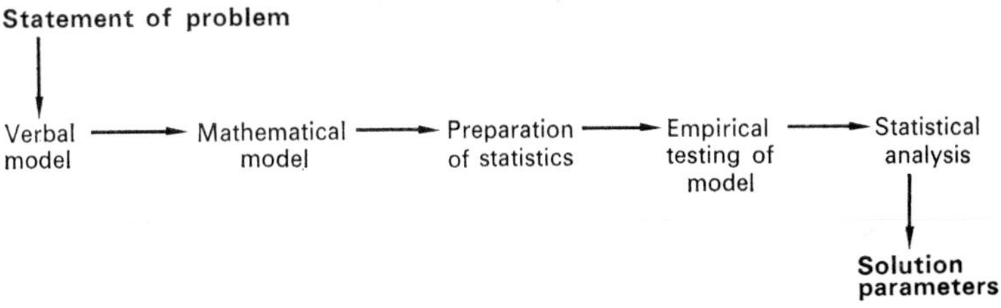

FIG. 5.5 Diagrammatic verbal model

equations in which the coefficients and other constants are expressed numerically. Perhaps the latter are best described as numero-mathematical models. Numerical models in the form of tabulations include profit and loss accounts, balance sheets and a wide range of statistical tables. Their forms and purposes are examined in Chapter 2.

The mathematical model

Mathematical models usually state an effect as a function of a cause or causes in an algebraic equation or in the form of a matrix or set of simultaneous equations. All verbal and numerical models concerned with cause and effect can be converted to mathematical models, but in most cases, there would be no profit in it. To be of practical value, a mathematical model must provide a stepping stone between hypothesis and statistical analysis.

Although many verbal models exist in their own right, the hypotheses underlying all numerical and mathematical models first see the light of day as verbal models. In all practical investigations, the problem comes first; then the verbal model; followed by the mathematical model, which must be tested for form and perhaps modified in the light of the available basic statistical data; statistical analysis; and finally the solution of the problem. Figure 5.5 sets out the chronicle of events as a verbal model in its own right.

Models in two variables

Mathematical models of the general form:

$$y=f(x)$$

are of no use at all for statistical analysis. They may be of value to pure mathematicians; to statistical analysts they convey no more than verbal models, 'y is a function of x'. What the statistician needs as his stepping stone from hypothesis to analysis is a model in the form of an algebraic equation of a particular type. Regression analysis can handle only linear equations, equations that can be converted to a linear form, and quadratics and equations of higher degree. The simplest linear equation expresses a variable, the dependent, as a function of one other, the independent variable:

$$y=ax+b \qquad \text{...(i)}$$

where y is the dependent variable; x, the independent variable; a, the coefficient of x; and b, a constant. Suppose a straw merchant charges £5 a ton for his merchandise, plus a delivery charge of £2 a consignment, then the total charge for a consignment of 10 tons would be:

		£
10 tons at £5	=	50
Delivery charge	=	2
Total		52

The charge per ton of £5 is equivalent to a in equation (i); the delivery charge of £2 to b; 10 tons to the independent variable, x; and the total charge of £52, to the dependent variable, y:

$$y = 5 \times 10 + 2 \text{ in £'s}$$

$$= £52$$

Many things in everyday life can be converted to the simple mathematical form of equation (i). Mathematicians with a natural flair for their subject tend to convert everything possible to a mathematical equation. The old formula for making a pot of tea, for instance, i.e. one spoonful of tea per person plus one for the pot becomes:

$$y = x + 1$$

where y is the total number of teaspoonfuls of tea required; and x, the number of persons. Another example is the domestic tariff used by some area electricity boards for raising electricity charges. There is a fixed charge per quarter, equivalent to b of equation (i) and a charge per unit consumed, equivalent to a, the number of units consumed in the quarter being represented by x.

In regression analysis, it is the average values of parameters a and b that are sought. Why then, it may be asked, do we not have two equations, one expressing a as a function of x, y and b; and the other, b, as a function of x, y and a, such for example as could be derived from equation (i):

$$ax = y - b$$

$$\therefore a = \frac{y - b}{x} \qquad \ldots \text{(ii)}$$

and

$$b = y - ax \qquad \ldots \text{(iii)}$$

As we shall see more clearly as we proceed, it is rarely if ever feasible to attack a problem of this kind so directly, especially where the parameters consist of an index or indices, and a multiplier:

$$y = x^a b \qquad \ldots \text{(iv)}$$

or the equation is a quadratic or is of higher degree:

$$y = b + a_1 x + a_2 x^2 + a_3 x^3 \ldots a_n x^n \qquad \ldots \text{(v)}$$

Regression analysis assumes there exists a series of x's and y's to enable the statistician to calculate the values of a and b in a two-variable problem. In practice, the value of a is calculated first and then it is used for solving for b. In a two-variable problem like that assumed to exist by equation (i), only two sets of observations are mathematically required to give a solution. If the following are two sets of observed values of x and y:

x	y
12	53
20	85

then by simultaneous equations or the calculus of finite differences, we have:

$$8a = 32$$

$$\therefore\ a = 4$$

and substituting 4 for a to solve for b, we have:

$$12 \times 4 + b = 53$$

$$\therefore\ b = 53 - 48$$

$$= 5$$

and the complete solution is:

$$y = 4x + 5$$

However, it is a far cry from this simple kind of textbook algebra to statistical analysis. Statisticians assume that their raw material is incomplete and subject to error, and they would never be satisfied with two sets of observations for any kind of analysis. They require what they call degrees of freedom, that is, some sets of observations in excess of the number of variables involved in the analysis, and they need at least ten; though sometimes they have no alternative to accepting a smaller number short of abandoning the investigation.

Finding the best-fitting model

We are concerned with mathematical model building in this chapter, not with statistical analysis. In the example above, we assumed that the appropriate model is linear, i.e., of the form of equation (i). Even from the purely mathematical point of view, a third set of observations is required to test the linear assumption and to decide the correct form of model. Suppose a third set of observations for the example reads $x=100$ and $y=377$. Then on the linear assumption and using this third set in conjunction with the second set above, we have:

Observations	x	y
Third set	100	377
Second set	20	85
Difference	80	292

i.e.

$$80a=292$$

$$\therefore a=3{\cdot}6$$

We therefore have

$$b=377-100\times 3{\cdot}6$$

$$=17$$

the complete equation being:

$$y=3{\cdot}6x+17$$

compared with the original equation:

$$y=4x+5$$

Clearly then the linear model is not satisfactory: it does not provide a good fit to the three sets of observations.

Let us try the formula of equation (iv), the logarithmic model:

$$y=x^a b$$

the linear form of which is:

$$\log y=a\log x+\log b \qquad \ldots \text{(vi)}$$

The logarithms of x and y are treated in exactly the same way as their natural numbers. Table 5.1 sets out the procedure in a convenient form.

TABLE 5.1 Testing the exponential model

x	$\log x$	*Difference* $\log x$	y	$\log y$	*Difference* $\log y$	$\frac{Dif.\ \log y}{Dif.\ \log x} = a$
12	1·0792	—	53	1·7243	—	—
20	1·3010	0·2218	85	1·9294	0·2051	0·925
100	2·0000	0·6990	377	2·5763	0·6469	0·925

The last column of the table provides all the evidence necessary to show that the best-fitting model for the example is logarithmic in form. Solving for b, we have:

$$\log b = \log y - a \log x$$

and using the first set of observations:

$$\log b = 1{\cdot}7243 - 0{\cdot}925 \times 1{\cdot}0792$$

$$= 0{\cdot}7260$$

$$\therefore\ b = 5{\cdot}32$$

and the complete equation is:

$$y = 5{\cdot}32x^{0{\cdot}925}$$

The complete equations are given as a matter of interest, and to demonstrate their arithmetic derivation. The values assigned to a and b are not averages subject to a margin of error: they are based on data which are accepted as perfect and complete. The statistical analyst never has perfect and complete data, and the values he would assign to a and b would be averages subject to a margin of error like any other average based on a sample.

Models in more than two variables

Multiple models, as models in three or more variables may be called, are much more common in practice than simple models, i.e. models in two variables. Time series are often the raw material of statisticians, and this fact alone introduces a third variable, time, into the model. But time itself can scarcely be a true factor: it is merely introduced as a catch-all device, which in itself is an admission that there are or may be factors besides the independent variables to take care of. That is one of the problems of statistical analysis. Indeed, the existence or possible existence of other factors which for one reason or another cannot be accounted for in the analysis, is the *raison d'être* of statistical analysis. If, for instance, all the factors in the demand for a product were known and measurable, the quantity of the product sold could be expressed as a function of those factors and all the coefficients determined by simple algebra without recourse to regression analysis. That would be a pure mathematician's dream. However, it would be quite impossible to account for all factors in the demand for any product. Suppose, for instance, the product in question is spaghetti, and that one Joe Scarlatti living at Clitheroe is a big consumer of spaghetti but has to reduce his consumption of the commodity by 50 per cent when he is thrown out of work. Then consider all the other people who find spaghetti to their taste and who may be in and out of work in turn. Their vicissitudes like Joe Scarlatti's would each form an independent variable in a complete demand model for spaghetti. Individually, they may not be significant; collectively, they could well be, and in case they have a trend in time, the time factor is introduced.

Statistical equations never take account of all factors, and for this reason they are referred to as stochastic. A stochastic equation is an incomplete equation as distinct from the complete equation of the pure mathematician and the textbook on algebra. It is not possible for a statistical analyst to fall into the error of supposing that his models are anything better than stochastic equations. However, there are many analysts who do not trouble about the shape of their models, but blindly adopt a linear form for all purposes. Some adopt a quadratic form for some purposes and a linear form for all others. How this unrealistic state of affairs comes about is a matter of conjecture. It is true that regression analysis can handle only linear equations, quadratics and equations of higher degree; but the list does not preclude curvilinear equations that can be converted to a linear form by the use of logarithms or other device.

Types of equation

An important attribute of a good mathematical model for statistical purposes is practicability. An equation that is partly exponential and partly linear is impracticable, e.g.,

$$y = x^a b + cz$$

which may have good theoretical grounds, but which cannot be converted to a linear form. In cases like this, the only solution is to assume that all factors, i.e., independent variables, bear the same or a similar relationship to the dependent variable as the most important one. Where z in this example is considered to be less important than x, then the appropriate form is logarithmic, the two alternatives being:

$$y = x^a z^c b \qquad \ldots \text{(vii)}$$

and:

$$y = x^a 10^{cz} b \qquad \ldots \text{(viii)}$$

Since the logarithm of 10 equals 1, the linear form of equation (viii) reads:

$$\log y = a \log x + cz + \log b$$

Mathematical statisticians use little epsilon, ϵ, the base of natural logarithms, instead of 10 in equations like (viii) above; but it does not affect the results. Ten is the more practical and is preferable on this account alone.

Where the basic data consist of time series, the factor *time* is introduced into a linear equation in the same way as other factors, e.g.;

$$y = ax + rt + b$$

where t is time and r is its coefficient. But in an exponential equation it is correctly introduced in the way that z is introduced into equation (viii) above, that is:

$$y = x^a 10^{rt} b$$

The reason for this may be explained by comparing z^c and 10^{cz} of equations (vii) and (viii) above as they would affect the value of y. Suppose the series of z consists of the numbers 1 to 10 and c is 0·1. The values of the two terms in the series are given in Table 5.2. Columns (3) and (5) show the figures added in respect of z to the right hand side of the two logarithmic equations. If the two series were plotted on rectilinear graph paper against the original series in column (1), 0·1 log z in column (3) would describe a curve rising to the right and concave downward, whereas 0·1 z in column (5) would describe a straight line rising to the right. The antilogarithms of the two series given in columns (4) and (6) are the figures which enter as multipliers in the exponential forms of the equations. The graph of the figures in column (4) is still a curve rising to the right and concave downward; but the graph of those in column (6) is concave upward.

One general conclusion that can be drawn from the illustration of Table 5.2 is that the higher values of z are given greater weight in equation (viii) than they are in equation (vii). It may be theoretically valid for a series representing time; but it is questionable whether it could be so for any series representing a true factor. In any event, apart from time as a factor, very careful thought should be given to any idea of introducing a factor into an exponential in the way that z is introduced into equation (viii).

TABLE 5.2 Comparison of z^c and 10^{cz}

Series z (1)	$\log z$ (2)	*0·1* $\log z$ (3)	$z^{\cdot 1}=$ *anti*-log *of* (3) (4)	*0·1* z (5)	$10^{\cdot 1z}=$ *anti*-log *of* (5) (6)
1	0·0000	0·0000	1·000	0·1	1·259
2	0·3010	0·0301	1·072	0·2	1·585
3	0·4771	0·0477	1·116	0·3	1·995
4	0·6021	0·0602	1·149	0·4	2·512
5	0·6990	0·0699	1·174	0·5	3·162
6	0·7782	0·0778	1·196	0·6	3·981
7	0·8451	0·0845	1·214	0·7	5·012
8	0·9031	0·0903	1·231	0·8	6·310
9	0·9542	0·0954	1·246	0·9	7·943
10	1·0000	0·1000	1·259	1·0	10·000

Another type of logarithmic equation may suggest itself. It is:

$$y=x^a czb \qquad \ldots \text{(ix)}$$

whose logarithmic form could be written:

$$\log y=a \log x+\log (cz)+\log b \qquad \ldots \text{(x)}$$

However, it is not practicable: the constant c and the variable z look to be attached to each other in these equations; but in practice they would have to be taken separately, since equation (x) wrongly supposes that there could be a basic series of statistics representing cz; c is one of the parameters that the analysis is intended to evaluate. The logarithmic form that might be suitable for analytical purposes would be:

$$\log y=a \log x+\log c+\log z+\log b$$

since $\log (cz)$ equals $\log c+\log z$. There would be confusion between $\log c$ and $\log b$. Indeed, since they are both constants, it would not be possible to keep them separate. Furthermore, the analysis would provide its own solution by rejecting c, and assigning a natural (non-log) coefficient to log z, thus converting the equation to the type of equation (vii), i.e.:

$$y=x^a z^c b$$

In practice, there is no room in any equation for more than one unattached constant; and every variable must have an attached coefficient or index.

Interdependence of factors

In the foregoing it appears to be assumed that of all the factors to be taken into account in an analysis, one can be selected as the dependent variable, all the others being independent. This kind of thing may hold good in some fields of enquiry; but in the economic and industrial field it is largely illusory. Independent factors are few and far between in the industrial and economic sphere. It is not only that factors react on one another: a change in A causes a change in B, which in turn causes a further change in A; it also happens in some cases that there is no theoretical clear cut dependence and independence of factors: from one angle A appears to cause B, from another, B appears to cause A.

However, in mathematical model building, it is necessary to express one of the factors as a function of the other factor or factors. Which factor should be chosen to serve as the apparent dependent variable can most often best be decided by reference to the problem which gave rise to the investigation, and the character of the solution that is being sought. Sometimes the necessary constant in a linear model clearly belongs to one factor and not another, and this must have some influence on the choice of the dependent variable. It could be argued that the rate of output depends upon the rate of input in terms of cost. Then we would have an equation like this:

$$Q=g(T-F)$$

where Q is, say, the annual rate of output; T, the total annual cost; and F, the annual fixed cost. The fixed cost is incurred in any event, so that input must be expressed in terms of the variable cost, i.e., $T-F$. But the object of the analysis may be to determine F as well as g, so that $T-F$ cannot be given expression in the form of a statistical series before the analysis is completed—a situation of deadlock. There is an alternative: to express input as a function of output. From the above equation, we have:

$$\frac{1}{g}Q = T - F$$

$$\therefore T = \frac{1}{g}Q + F$$

Since $1/g$ is just as much a constant as g, we can rewrite the equation:

$$T = aQ + F$$

where $a = 1/g =$ the marginal cost, which, as it happens, is more meaningful than its reciprocal, g. The analytical statistician would have no difficulty with a model like this provided he had enough sets of observations of T and Q; F, of course, is the constant. As indicated above, this is the accepted standard form of cost model for a single brand of product.

The basic data

Another attribute of a satisfactory realistic model designed for statistical analysis is conformity to the basic statistical data available. There would be no point in including in a model variables for which no basic data exist.

Data for internal factors are usually available, though not always in the form best suited for solving the problem in hand. It is the data for external or environmental factors that give rise to the greatest difficulties. All firms, even the state monopolies, trade in a competitive environment; and some environmental factors are of great importance in marketing and market research. One such factor is the trend in the price-level of brands in direct competition with the firm's brand or brands. Export markets present their own peculiar problems: changes in import duties and in freight rates, for instance, which affect distribution costs.

A preliminary examination of the basic data available or that can be made available is an essential step in the model-building process. It is not often possible to reshape basic data to suit a model; the model must be shaped to suit the basic data. If a model thus shaped would not provide a solution to the problem giving rise to the analysis, the investigation may have to be abandoned. Conflicts of this kind are not uncommon; for available data are not always ideally defined for the purpose.

Much depends upon the nature of the problem to be solved. Where the object is to enable the management accountant or the market research department to estimate or forecast the value of y for planned or programmed values of x, z, etc., perfection of data definition is less important than where the object is to obtain parametric values such as the price-elasticity of demand for a brand of product, a marginal cost or the average time taken to perform a certain task in circumstances where direct measurement would be difficult. Parametric values are, of course, necessary for forecasting the value of y: but since their use is internal to the model, their precise definition is unimportant. It is where the parameters themselves are to be used externally, as the price-elasticity of demand and marginal cost are for rational pricing, that the definitions of solution parameters and therefore of basic data are important.

Absolute values generally provide the most suitable form of basic data for analytical purposes. The conversion of absolute values to ratios for such an arbitrary reason as reducing the number of variables in the analysis is to be avoided. Some basic data are unavoidably expressed in ratio form: price, for instance, as covering a period is the ratio of the sales proceeds to the quantity sold; it provides the major independent variable in the analysis of market demand. Again, the real price may be derived from the sales proceeds deflated by reference to the retail price index: i.e. from the real sales proceeds, which is the ratio of the money sales-proceeds to the retail price index. The money price of a product can be corrected for changes in the purchasing power of money directly, by taking the ratio of the money price to the retail price index—the ratio of a ratio. Some statisticians prefer to introduce the retail price index into a market demand analysis in the form of an additional independent variable, an approach that appears to have a number of disadvantages over the alternative of using the index as a means of correcting prices and other value data for changes in the purchasing power of money:

(*a*) It necessitates what seems to be a meaningless and useless concept: the elasticity of demand with respect to the retail price index;

(*b*) There seems to be a good case for purifying basic data for such a disturbing factor rather than regarding it as a true factor;

(*c*) It is bound to distort the solution values of such useful parameters as the elasticity of demand with respect to (i) the price of the product; (ii) total personal income; (iii) the firm's expenditure on advertising and sales promotion, (iv) competitors' expenditure on advertising, etc; and so on: and

(*d*) It seems to be an easy way round the difficulties created by changes in the purchasing power of money. Yet changes in the purchasing power of money can influence the demand for a product or brand only through their effect on price. If the money price of a brand remains constant while the purchasing power of money falls by 10 per cent, the effect on demand will be much the same as the effect of a 10 per cent fall in money price at a time when the purchasing power of money remains constant. At any rate, that is the assumption, which seems a safe and valid one, of those statisticians who prefer to deflate prices and other values embraced by the demand model by reference to the retail price index.

There is a strong temptation to use a dependent variable in terms of a ratio: miles per gallon, output per man-hour, cost per ton, and so on. Ratios of this kind are useful in planning and control; and a list of the factors that create changes in them is equally useful. But is it possible to measure the effect of the factors? And if so, would it be worth while? Consider, for instance, vehicle fuel-consumption: m.p.g. may be regarded as a function of m.p.h. and miles per journey, and a linear form of model for analytical purposes would be:

$$\frac{M}{G}=a\frac{M}{H}+b\frac{M}{J}+K \qquad \ldots \text{(xi)}$$

This brings into the light of day a paradox: we are concerned with fuel consumption, not vehicle mileage except as a factor in fuel consumption. Gallons per mile as the dependent variable would be more sensible from this point of view. This being so, then it seems that we could regard total fuel consumption as the dependent variable, and total vehicle mileage as an independent variable along with the total hours in traffic and the total number of journeys:

$$G=aM+bH+cJ+K \qquad \ldots \text{(xii)}$$

in which a, the coefficient of M, represents the gallons per mile independently of the hours in traffic, H, and the number of journeys, J.

There may also be two more useful independent variables involved, viz.: total loads carried in terms of weight, and the cumulative vehicle mileage run since the last engine overhaul or major service. Both these would often have a significant effect on fuel consumption. Needless to say, apart from the cumulative vehicle mileage, all the variables including the total loads carried would apply to some unit of time, perhaps a week or month or year, and each set of observations could relate to a different vehicle, or to the same vehicle or vehicles over a period. If the latter, it may be considered advisable to introduce time as a factor.

Some basic mathematical models

There is probably no limit to the number of models that could be of use to management accountants. Every industry including banking, insurance, stock-broking, electricity and other service industries as well as the extractive and manufacturing industries has its own peculiar requirements. And every basic model has its variants to satisfy such requirements. The following paragraphs offer and discuss a few basic models that have been found appropriate and realistic in practice. With the possible exception of the cost model, the equations indicated may not hold good in all situations, and it is suggested they should be tested for type by means of a suitable one of the regression techniques referred to and demonstrated in Chapter 7, before they are applied to a particular analysis. The analyst should not adopt them blindly, but should regard them as a pointer with respect to both type of equation and variables included.

The annual cost model

Annual cost models are nearly all linear in type. For a single product it is:

$$T = aQ + F \qquad \ldots \text{(xiii)}$$

where T is the total cost; a, the marginal cost; Q, the quantity produced; and F, the annual fixed cost. For a plurality of products, it is:

$$T = a_1Q_1 + a_2Q_2 + a_3Q_3 \ldots + F \qquad \ldots \text{(xiv)}$$

where subscripts 1, 2, 3 ... indicate different products or brands. It may be possible on rare occasions to convert equation (xiv) to a more sophisticated form, which attempts to allocate part of the fixed costs to the individual products:

$$T = a_1Q_1 + b_1N_1 + a_2Q_2 + b_2N_2 \ldots + F_h \qquad \ldots \text{(xv)}$$

where b represents the annual fixed cost of a unit of capacity particular to the product; N, the number of units of particular capacity available for use; and subscript h denotes the hard core of fixed costs, i.e., those fixed costs that cannot be allocated to individual products except on an arbitrary basis. Undoubtedly this equation is somewhat theoretical in conception: it assumes that a product is produced in a plant that consists of a number of units of equal capacity and each incurring the same additional annual fixed cost, which would consist for the most part of annual capital charges and indirect labour. Just as a is the marginal cost with respect to the product, so b is the marginal cost with respect to plant capacity, bN being the variable cost with respect to capacity, but fixed with respect to product. In the equation, N_1, N_2, N_3 . . . are variables. It is important to note that if any of them showed no variation in the sets of observations available, then one or more bN's would be constants and the regression analysis would therefore merge them with F_h, the unattached constant. All variables must vary if they are to emerge together with their coefficients from a regression analysis, and not merely form part of the constant. The same applies to a pure mathematical solution by simultaneous equations, in which the elimination of the constant also eliminates any variable that does not show a different value in at least one set of observations.

The annual demand model

There does not seem to be any practicable alternative to the exponential type of model expressing quantity sold in unit time as a function of price and other factors. Both theory and practical experience indicate that the exponential type is appropriate:

$$Q = Kp^e s^f R^g D^h E^j \qquad \text{. . . (xvi)}$$

where Q is the quantity of the product or brand sold; p, the price of the product deflated for changes in the purchasing power of money; s, the price index of competitors' brands similarly deflated; R, the firm's own expenditure on advertising the product, also deflated; D, competitors' estimated expenditure on advertising their brands (deflated); and E, the total available income of consumers (e.g. personal income, from the *National Income* blue book) (deflated); and the indices e, f, g, h, j represent the elasticities of demand with respect to the factors. The price-elasticity, e, determines the slope of the sales proceeds curve. The other elasticities affect its elevation. Chapter 8 discusses in greater detail the type of model appropriate to demand analysis.

The annual production model

Here, the ground is less firm. Theory has little or nothing to say, though admittedly, economists have been talking about the *production function* for many years. So far as has been made public, there has been little experience in deriving production models for analytical purposes, but what there is indicates the exponential form[1]:

$$Q = KF^a W^b E^c \qquad \ldots \text{(xvii)}$$

where Q is the annual output or index of output; F, the average labour force; W, the average weight or size of the products; E, the past experience on the products as measured by the cumulative output; T, the average number of types or brands in production during the year; and a, b, and c, are the parameters whose values the analyst is seeking to determine. The variable W applies only where the products consist of manufactured assembled articles such as cars, aeroplanes, bicycles, pieces of furniture, washing machines, lathes, printing machines, typewriters, and so on. The variable E, the cumulative output, is included to take care of the learning potential of the labour force. It suggests another variable for inclusion, i.e., a measure of the rate of labour turnover in the factory or factories, especially where the rate changes significantly in the period covered by the sets of observations to be used for the analysis. If the learning potential is significant, an increase in the rate of labour turnover would have the effect of reducing, and a decrease of enhancing, its effect on production.

The labour-time model

There is no reason for supposing that a model designed for the analysis of labour time is anything but linear. Indeed, it is questionable whether any other type could give the kind of information that is likely to be required. As to the variables that are to be included, each investigation would have to be considered on its merits. One labour-time model that has been applied in solving a practical problem regarding a parcels collection and delivery service, is:

$$H = aM + bE + cP + K \qquad \ldots \text{(xviii)}$$

where H represents the total man-hours in a working day or week actually spent on the work; M, the mileage covered; E, the number of effective calls made; P, the number of parcels or consignments dealt with; and K is a constant whose value would in theory at least be the average time spent daily or weekly on preparing the vehicle for the road, meal times and the like. Each parametric value would represent the average time taken, a to travel a mile, b to make an effective call,

[1] See, for example, E. J. Broster, 'Productivity in the wartime aircraft industry', *Aircraft Engineering*, June 1957.

and c to handle a parcel at the terminal. The average number of journeys daily or weekly might provide an additional independent variable, or it might be considered more relevant to the analysis than the number of parcels or consignments handled.

Regression analysis depends for its successful application upon variation of the variables, each of which must vary within itself, and also differ in degree of variation from each of the other variables. It can handle cases where one variable varies as the square or square root of another variable, but not where two variables vary *pro rata* one to the other. This is particularly relevant to the analysis of labour time; for in manufacturing industry, the number of times each process is applied usually has a *pro-rata* relationship to other processes. It would therefore not be possible as a rule to analyse labour time on different processes by regression analysis. However, a mathematical model might be useful for costing, for instance, but the parametric values would have to be determined by a work-study method.

Other models

That there are many other mathematical business models that could be used in statistical analysis has already been pointed out. It will be seen from the above that there are two distinct kinds of model. First, there are those models whose coefficients are expressed in the same terms as the dependent variable. The models of annual cost and labour time are of this kind. It is difficult to see how the variables times their coefficients could be anything but additive, their sum together with the constant giving the value of the dependent variable for any given values of the independent variables. In other words, the type of model that satisfies this kind of analysis is necessarily linear, or very rarely, if ever, of quadratic form.

Secondly, there are those models whose coefficients would not, even if they were assumed to be linear, be expressed in the same terms as the dependent variable. Their variables times their coefficients would not necessarily be additive and the best-fitting type of equation may be logarithmic, especially where an increase in any variable would, other things being equal, be expected to cause a decrease in the dependent variable and *vice versa*. Demand for instance is a function of price amongst other factors; and it varies inversely with price. If we assumed a linear type of model, we would have:

$$Q=K-ap$$

which seems to make nonsense, since anything multiplied by price would be expressed in terms of value not of quantity. It does not necessarily follow that the best fitting type of model is anything but linear; but the equation does suggest that a linear type found empirically should be treated with reserve and the model retested for linearity later when enough additional basic data have become available.

The production model can be examined in the same way. The major independent variable is the labour force. Here, given linearity, its coefficient would be expressed in the same terms as the dependent variable, the quantity produced. However, that would mean that production varies *pro rata* to the labour force, whereas it has been found in practice, that is, empirically, that it tends to vary logarithmically owing to the economies of scale. In addition, where a learning law operates, there seems to be no alternative to the logarithmic type of model.

Two or three methods of regression analysis can be used for determining empirically the types of models. They are considered and demonstrated in Chapter 7.

6

Statistical correlation

Of all the statistical techniques that have ever been expounded and applied to practical problems, statistical correlation is the most misused. It is also one of the least useful, or perhaps it would be more accurate to say that it is not half so useful as many statisticians seem to think.

Statistical correlation is not to be confused with regression analysis, which used to be commonly called correlation analysis. Statistical correlation is concerned with the measurement of the statistical relationship between two or more series of statistics; regression analysis, with the statistical measurement of the effect that a factor or each of more than one factor appears to have on another factor.

If series A tends to rise when B rises and to fall when it falls, A and B are said to be positively correlated. If A tends to fall when B rises, and to rise when B falls, they are said to be negatively correlated. There is a numerical measure of correlation; it is called the coefficient of correlation, or, rarely nowadays, the Pearsonian coefficient of correlation after Karl Pearson, who first developed and expounded it. It ranges from $+1{\cdot}0$ for perfect positive correlation to $-1{\cdot}0$ for perfect negative correlation; zero indicates there is no correlation at all. In practice, there can be perfect or near perfect correlation, positive or negative, but rarely if ever zero correlation. Two arithmetic progressions, e.g. 1, 2, 3 . . . and 20, 25, 30, 35 . . ., are perfectly correlated positively. One turned upside down, e.g. 1, 2, 3 . . . 100, 95, 90, 85 . . . and they are still perfectly correlated, but this time negatively. Since two series of random numbers would almost invariably reveal a degree of correlation between them, it would not be easy to offer two series that would have a correlation coefficient equal to zero.

In economic and industrial statistics, some degree of correlation probably exists between any two series through a chain of cause and effect that it would often be impossible to unravel. The shorter the chain the greater the coefficient of correlation would tend to be.

USES OF CORRELATION

It was stated in Chapter 5 that in the economic and industrial field all models or equations expressing one factor as a function of another or others for statistical purposes are stochastic, i.e., they are incomplete in the sense that some independent variables are necessarily omitted either because they are unknown or because there is no satisfactory numerical measure of them. Even where the statistical observations are perfectly accurate, it follows that figures for the dependent variable as compiled from the complete equation determined by regression analysis will nearly always differ from the original series for the dependent variable used in the analysis.

Nine times out of ten there is no reason for supposing that two causally related factors as such are not perfectly correlated. That is, if we had accurate statistics for a period or situation in which all other factors remained dormant, the two series would prove to be perfectly correlated. But our statistics are rarely accurate or of the right definition, and other factors are never dormant. Imperfect materials will never yield perfect results. It is the materials, the statistical data, that are being correlated not the factors themselves. To correlate factors, only logic and theory can be applied. In short, the correlation coefficient gives a measure of the correlation between series of figures representing factors, not between the factors themselves. Statistics and factors are like words and deeds: there is a world of difference between them. It is as important to distinguish between factors and their statistics as it is between things and their names.

MISUSES OF CORRELATION

Statistics, it is often said, can be made to prove anything. It is when statistics are misused that they can be made to prove anything. Statistical correlation lends itself to misuse, and quite innocent misuse at that. People using the technique can very easily mislead themselves into believing something that is quite untrue.

Self-deception results mainly from the misinterpretation of statistics. The interpretation of statistics, often referred to as statistical inference, is strictly a logical process. Statistics are facts expressed in numerical terms, and the thought processes needed in statistical inference are precisely the same as those brought into play when one draws inferences from non-numerical facts. These thought processes come into play, or ought to do, in statistical correlation, as much as in any other sphere of statistics.

It is outside the regression field where the difficulties and dangers lie. Here the analyst is concerned with cause and effect. The most common error is the one known in logic as the false cause or *non causa pro causa*. It is a kind of reasoning that is frequently found in the bar parlour. Much of our weather lore is based on it: red at night, shepherd's delight; red in the morning, shepherd's warning; rain before seven, fine by eleven. The attribution of a change in the weather to the passing of the full moon is more particularly a fault of logic known as *post hoc ergo propter hoc*—coming after, therefore a result of, which is a form of the false cause.

When two things, A and B, happen simultaneously, there is a strong temptation to attribute one to the other. Often, of course, there is evidence extraneous to the events to support the attribution. If a rifle is fired at a formal target and a hole appears in the bull's-eye, it is fairly safe to attribute the hole to the firing of the rifle. Here the real evidence is extraneous to the events: it consists of a knowledge about the use of rifles and the skill of the marksman.

For A and B to happen simultaneously, there are six possible explanations:

1. A is wholly due to B;
2. B is wholly due to A;
3. A is partly due to B;
4. B is partly due to A;
5. A and B are joint effects of an extraneous cause;
6. The simultaneity of A and B is a coincidence.

In statistical correlation, it is safe to rule out the last of these explanations provided there is a sufficient number of observations to show that A and B nearly always happen simultaneously. When they happen simultaneously always, there is no need for a statistical analysis to prove the existence of a causal relationship between the two. That is all the statistician can prove: that there is a causal relationship between A and B. Contrary to a fairly common belief, he cannot prove that there is no causal relationship, for the absence of a significant degree of statistical correlation between two series of figures representing A and B does not prove that they are not causally related. As shown below, extraneous factors may have a highly disturbing effect on the statistics of the effect. An expression commonly found in the literature is *spurious correlation.* Its opposite, *spurious non-correlation,* seems to be a concept of equal importance. At any rate, it would bring home to the student of statistics and the statistician that the absence of correlation between two statistical series proves precisely nothing.

Whatever statistical correlation may be capable of doing, it cannot prove the nature of a relationship between two factors, whether A causes B, or B causes A, with or without extraneous factors, nor can it help in deciding whether A and B are joint effects of another factor. Underlying any correlation exercise, there must be a hypothesis derived by deductive reasoning. Where evidence external to the statistics exists that A causes B, or where A always happens when B happens and *vice versa,* statistical correlation can serve no useful purpose: a causal relationship is already established. But these kinds of phenomena are of no interest to the statistician; he enters the field where other evidence is inconclusive or speculatively theoretical, and where A can happen without B or B can happen without A, or both. It is very difficult ground. Logically, all the statistician can do is to show that one event is more likely to take place when the other event takes place than when it does not.

If A always happens when B happens and never at any other time, logicians call B *a necessary and sufficient condition* to A. In such circumstances a statistical correlation would be superfluous. It is only when one or the other event sometimes takes place without the other's taking place that statistical correlation can be of any service. The statistician may be on dangerous ground, he may know he is on dangerous ground; but statistical techniques are often the only way available for showing the existence of a causal relationship. Even so, a negative result, i.e., a low degree of correlation between two series, is no proof that the two factors are not in fact closely causally related. A significant result, i.e., a high degree of correlation, is the only acceptable evidence.

In the lung-cancer-cigarette-smoking investigations, enough medical evidence to prove a causal relationship would have rendered the statistical investigations entirely superfluous. Medical evidence would have shown what is cause and what effect; and this the statistical evidence does not do: it merely proves the existence of a causal relationship. The investigations began in a highly controversial atmosphere about 1950 when Drs Doll and Bradford Hill reported their findings on the subject in the *British Medical Journal* and inferred from a significant degree of correlation between the available statistics of cigarette consumption and the incidence of lung cancer that cigarette smoking caused lung cancer. The statistical evidence did not justify any such inference; all it did at most was to demonstrate the existence of a causal relationship.

Since many heavy smokers of cigarettes live long and die of old age, cigarette smoking is not a sufficient condition; and since there are many people who have never smoked but nevertheless contract lung cancer, it is not a necessary condition either. There must be some other causal factors, but the statistical investigations do not give a clue to their identity.

It is the purpose of these paragraphs not to join sides in a controversy that the British medical authorities appear to have settled at any rate to their own satisfaction, but to demonstrate by recruiting a widely known real-life example the pitfalls that lie in the path of those who seek to prove what is cause and what effect by statistical correlation. Sufficient to conclude by saying that although it may seem absurd to suggest that lung cancer might cause cigarette smoking, there is a possibility that whatever in the human body causes lung cancer also creates a craving for cigarette smoke. In the same way, there is a possibility that the two phenomena are joint effects of some extraneous cause. Either of these would explain the high degree of correlation found by Drs Doll and Bradford Hill, who, indeed, did not overlook their possibility but dismissed them in a short paragraph.

STATISTICS IN DEDUCTIVE REASONING

None of this is to say that statistical evidence is never of any use for showing the nature of a causal relationship. But it is necessary to use it in applied deduction, whereas statistical correlation is applied induction. In economics, both micro and macro, there are many instances of deductive reasoning which call on statistics for evidence for determining what is cause and what effect. For instance, the supply of and demand for a commodity in a free market are highly correlated, and their measures for farm and horticultural products in particular are the same series, i.e., the quantity changing hands. The balance is brought about by the price mechanism. A decrease in the demand or an increase in the supply causes a fall in the price and *vice versa*; and the change in the price preserves the balance between supply and demand. When a change takes place in the quantity changing hands in any period compared with the preceding period, an examination of the statistics of prices relative to that change would indicate whether it was a movement in the supply position or in the demand position that had initiated the change. In short the law of supply and demand is applied to the relevant statistics deductively.

THE COEFFICIENT OF CORRELATION

Sir Francis Galton, who flourished in the mid-nineteenth century, discovered a statistical correlation between intellectual distinction and family, and came to the conclusion that intelligence is largely inherited. However, the modern idea is that intellectual distinction depends upon childhood environment as much as or more than it depends upon heredity. Intelligent parents tend to provide a more intellectual environment for their children than unintelligent parents do.

Galton did some early useful work on statistical correlation. He used the graphic method and personal judgment. He plotted one series measured on the vertical axis against the other measured on the horizontal axis. If the dots thus plotted fell closely into alignment, there was a high degree of correlation between the two series. If they scattered far and wide, there was no correlation worth speaking of. The dot diagram, sometimes called the scatter diagram and occasionally the Galton graph, still remains a useful device for judging correlation in a single exercise. Where comparison between two correlations is required, something more precise than personal judgment is called for. Karl Pearson answered the call, and so we now have the more precise measure of correlation. The formula of the coefficient of correlation, which is usually denoted by r, is as follows:

$$r = \frac{\Sigma(xy)}{\sqrt{\Sigma x^2 . \Sigma y^2}} \qquad \ldots \text{(i)}$$

where x and y represent the deviations from their means of two series X and Y being correlated.

Although it is necessary to use a large number of sets of observations, at least 20, to obtain significant results, a small number can be used for demonstration purposes. Table 6.1 contains a simple demonstration of the processes involved in calculating the regression co-efficient. The correction sums are arrived at in the way described in Chapter 3,

TABLE 6.1 Calculating the correlation coefficient for X and Y

	Series				
	X	Y	X^2	Y^2	YX
	1	0	1	0	0
	2	4	4	16	8
	3	7	9	49	21
	4	6	16	36	24
	5	8	25	64	40
Totals	15	25	55	165	93
Correction sums			45	125	75
Sums of squares and products of deviation			10	40	18

$$r = \frac{18}{\sqrt{10 \times 40}} = +0{\cdot}90$$

the sums of squares and products of the deviations being derived as follows:

$$\sum x^2 = \sum X^2 - \frac{(\sum X)^2}{n}$$

$$\sum y^2 = \sum Y^2 - \frac{(\sum Y)^2}{n}$$

$$\sum(xy) = \sum(XY) - \frac{\sum X . \sum Y}{n}$$

where n is the number of sets of observations.

In the example of Table 6.1, the two series X and Y are positively correlated, the coefficient being $+0{\cdot}90$. It is important that the sign should always be shown. Where the correlation is negative, the correcting sum $(\sum X . \sum Y)/n$, exceeds $\sum(XY)$. Suppose the Y series in Table 6.1 had been in reverse order, then $\sum(xy)$ would be -13 as shown in Table 6.2.

It will be noticed that the formula of the correlation coefficient, equation (i) above, is symmetrical in the sense that the two series play equal parts in it. The symmetry taken in conjunction with the fact that the numerator can take either a plus or a minus sign, makes it easy to remember.

TABLE 6.2 Negative correlation

	X	Y	XY
	1	8	8
	2	6	12
	3	7	21
	4	4	16
	5	0	5
Totals	15	25	62
Correction sum			75
Sum of products of deviations			−13

$$r = \frac{-13}{\sqrt{10 \times 40}} = -0{\cdot}65$$

Limits of the correlation coefficient

In non-regression problems, the formula of the correlation coefficient has to be applied somewhat blindly direct to the basic data. Not that much harm can result from such a procedure; for the solution coefficient will never exceed the true coefficient except by the kind of chance that exists in any circumstances; but it might be lower sometimes.

The fact of the matter is that the formula of equation (i) above can give the true coefficient only where the regression relationship between the two series is linear, and this proposition applies equally to regression and non-regression problems. Where the regression relationship is not required there is not much point in going to the trouble of finding the type of model that best satisfies it, except where theory clearly points to a particular type. There is, of course, no reason why the formula of the correlation coefficient should not be applied to any pair of series so long as it is understood that the solution is based on the assumption that the regression relationship is linear. An example of a known non-linear relationship in which the correlation is perfect is given in Table 6.3. The solution coefficient of $+0{\cdot}98$ is high by any standard; but since the two series are perfectly correlated, the true coefficient is $+1{\cdot}00$. Since the relationship can be converted to a linear form:

$$\log Y = 2 \log X$$

the true coefficient can be obtained by correlating the logarithms of

TABLE 6.3 A perfect non-linear relationship, $Y=X^2$

	X	Y	X^2	Y^2	XY
	1	1	1	1	1
	2	4	4	16	8
	3	9	9	81	27
	4	16	16	256	64
	5	25	25	625	125
Totals	15	55	55	979	225
Correction sums			45	605	165
Sums of squares etc. of deviations			10	374	60

$$r=\frac{60}{\sqrt{3740}}=+0{\cdot}98$$

the two basic series. An example of a perfect linear relationship in which $Y=2X$ is given in Table 6.4. A mere glance at the two series is enough to confirm the figure of $+1{\cdot}00$, for both are simple arithmetic progressions. Two geometric progressions, too, would give a coefficient of $+1{\cdot}00$ provided they had the same common ratio, since their relationship would likewise be linear.

TABLE 6.4 A perfect linear relationship, $Y=2X$

	X	Y	X^2	Y^2	XY
	1	2	1	4	2
	2	4	4	16	8
	3	6	9	36	18
	4	8	16	64	32
	5	10	25	100	50
Totals:	15	30	55	220	110
Correction sums			45	180	90
Sums of squares etc. of deviations			10	40	20

$$r=\frac{20}{\sqrt{400}}=+1{\cdot}00$$

Influence of other factors

It is emphasised above that a low correlation coefficient is not a sure sign that two factors are not causally related. Consider, for instance, the X and Y series in Table 6.5. The coefficient is very low, so low indeed that it can be said there is no correlation at all between the

TABLE 6.5 An example of a misleadingly low statistical correlation

	X	Y	X^2	Y^2	XY
	1	2	1	4	2
	2	6	4	36	12
	3	10	9	100	30
	4	4	16	16	16
	5	4	25	16	20
Totals	15	26	55	172	80
Corrections			45	135·2	78
Sums of squares etc. of deviations			10	36·8	2

$$r=\frac{2}{\sqrt{368}}=+0{\cdot}10$$

two series. Now suppose another factor, Z, is involved as an independent variable along with X, its series being as shown in the first column of Table 6.6. Also suppose (without anticipating Chapter 7) that it has been found by regression analysis that the model is of linear form and that the solution equation is:

$$Y=2X-2Z+8$$

Table 6.6 restates the Y series, and the next column shows the values of $-2Z+8$, which are deducted from Y to leave the effect on it of X only. This corrected Y series is given in the last column, and it will be seen that it is an arithmetic progression like the X series in the first column of Table 6.5. As factors, then, X and Y are perfectly correlated, in spite of there being a very low degree of correlation between their basic statistical series.

TABLE 6.6 Eliminating Z and the constant from Y

Z	Y	$-2Z+8$	$Y-(-2Z+8)$
4	2	0	2
3	6	+2	4
2	10	+4	6
6	4	−4	8
7	4	−6	10

It might be thought that the same kind of argument could be logically applied to cases of high correlation, that a high correlation between two series may be due to the effect of other factors. It could be true in the sense that one interpretation of a correlation coefficient is that it gives a measure of the extent to which variations in the dependent variable are accounted for by the independent variable, the amount by which it falls short of 1·00 being accounted for by other factors. If there are two independent variables that account for all variations in the dependent variable then if they happen to be highly and positively correlated themselves, either of the two will be highly correlated with the dependent variable. Such a possibility could make nonsense of the interpretation in question. The coefficient as calculated would be spurious from this point of view.

There is another point. In the industrial and economic sphere, an interpretation that assumes that there is a clear-cut dependent variable is usually out of the question. It is worth recalling in this connection that the formula of the correlation coefficient as given in equation (i) above is symmetrical. Statistical correlations are carried out for testing causal relationships, and for discovering whether there would be a danger of spurious correlation in a projected regression analysis.

Small samples

As the calculated correlation coefficient based on a sample or a small number of observations, tends to exceed the correlation existing in the universe as a whole, a correction to the calculated figure is often worthwhile. The correction formula is quite simple:

$$\bar{r}^2 = 1-(1-r^2)\left(\frac{n-1}{n-2}\right)$$

where $\bar{r}$ is the corrected coefficient, and n is the number of observations.

Where $n=5$ as in the worked examples above, then for $r=+0{\cdot}60$, we have

$$\bar{r}^2=1-(1-0{\cdot}36)\left(\frac{4}{3}\right)$$

$$=1-0{\cdot}85$$

$$=0{\cdot}15$$

$$\therefore \quad \bar{r}=+0{\cdot}39$$

and for

$$r=+1{\cdot}0$$

$$\bar{r}^2=1-\frac{4}{3}(1-1{\cdot}0)$$

$$=1-0$$

$$\therefore \quad \bar{r}=+1{\cdot}0$$

It seems, then, that the formula assumes there can be no sampling error in a correlation coefficient of 1·0. Where $r=+0{\cdot}90$, $\bar{r}=+0{\cdot}87$, for $n=5$. A general inference is that the closer the coefficient approaches to zero, the greater is the proportional sampling error. There is another general inference: since $(n-1)/(n-2)$ is smaller for large values of n than for small values, the larger the sample, the smaller is the proportional sampling error. It is shown above that where $n=5$ and $r=+0{\cdot}60$, then $\bar{r}=0{\cdot}39$. Where $n=10$, then for $r=0{\cdot}60$, the value of $\bar{r}$ is $+0{\cdot}53$.

In the process of correction the sign is lost. Strictly, the figures for $\bar{r}$ given above should be shown as plus or minus, since they are square roots. But where the sign is important as it usually is, that assigned to $\bar{r}$ should be the same as that of r. One more point: if the value of $\bar{r}^2$ turns out to be zero or a negative quantity, the value of $\bar{r}$ should be taken as zero.

Significance

As stated earlier in this chapter, any given value of the correlation coefficient is the more significant the greater the number of observations on which it is based. There are shades of significance; but the mathematical statisticians have provided figures of the minimum values of the coefficient for the correlation to be likely rather than unlikely for different numbers of observations or sizes of sample:

Size of sample	*Minimum acceptable values of r*
5	0·88
10	0·64
15	0·52
20	0·45
40	0·22
100	0·20

All the worked examples given above in the numbered tables are for five sets of observations. For this, the minimum acceptable value of the coefficient is 0·88, which seems rather high. However, the real trouble is that five as the number of sets of observations is much too small. Five sets have been used, and the formula applied to figures consisting of one or two digits only to make the examples easy to follow. It seems to be unnecessary always to lend verisimilitude to examples worked for demonstrating method by using real-life statistics or by giving a real-life name to the statistical series employed in the examples. Sufficient if it is understood that X and Y represent sets of observations of two related variables.

7

Regression analysis

It is scarcely possible in a single chapter to do full justice to a subject as big as regression analysis. However, it is intended here to do rather more than whet the appetite. It is intended to help the management accountant to an understanding of the practical elements of this valuable *open sesame* to the treasures of management information hidden away in company routine statistics. In common with other statistical methods, regression analysis has been successfully applied to many problems in a wide range of fields, from psychology to finance, from agriculture to medicine, and from economics to chemistry. There is reason for supposing that regression techniques can be applied with equal success to the field of management accounting and market research more extensively than they seem to have been applied in the past.

STATING THE PROBLEM

Statistical methods are no more than tools for the use of investigators. Apart from practice by students and the presentation of examples by writers and teachers, there is little profit in applying regression analysis to two or more series of statistics of related factors merely to see what happens. In practice, there must be a problem requiring solution; in short, the problem comes first. Needless to say, it should be stated clearly and as precisely as possible; and the investigator must keep it in mind throughout his work of seeking a solution. The problem can be stated as an objective expressed in the form of a question, e.g. 'What is the marginal cost of production of our typewriter model 3126?' 'How long on average do our road tanker drivers spend on making a collection (of milk from a dairy farm) or a delivery (of petrol or diesel oil to a service station or other premises)?' 'To what extent does a given change in our routine advertising affect sales?' 'What is the price-elasticity of demand for our typewriter model 3126?' It is not suggested that regression analysis will always give an answer. Everything depends upon the basic statistical data available or that can be collected and collated within the time allowed and the budgeted expenditure on the project.

FORMULATING THE HYPOTHESIS

The next step is to formulate the appropriate hypothesis in the light of the basic statistics available or that can be made available. It will call for the logical analysis of the objective and some deductive reasoning applied to the theoretical and practical evidence available, including the work of other investigators in the same or similar fields. The hypothesis will ultimately resolve itself into a verbal model listing all the relevant factors and showing which is considered to be the most convenient or the most suitable to adopt as the dependent variable, and of the rest, which is considered to be the most important as an independent variable in the light of what is judged to be its relative strength and of the object of the exercise. The translation of the verbal model into a mathematical model is discussed and demonstrated in Chapter 5.

THE BASIC STATISTICS

A preliminary inspection of the statistical material available should be made before any attempt is made to formulate a hypothesis and build a model; for there would be no point in considering a regression analysis let alone preparing for one unless suitable raw material were available. Just as the little girl in *French Without Tears* discovered there is no good substitute for flour and sugar for making a cake, so the statistical analyst will discover there is no substitute for good basic statistics to serve as the ingredients of an analysis.

The use of indicator material such as the volume of newsprint consumed to indicate the volume of output of newspapers, as the CSO does for the official index of production, calls for great care in a regression analysis. Indicator material can be better than nothing in making an index number of prices or production; it can be worse than useless for an exercise covering so narrow a field as a regression analysis. It is advisable never to use indicator material for major factors, and to use it only with great care for minor factors.

Whether to use absolute values or ratios will have been decided at the model-building state; and now is the time to prepare the basic statistics in accordance with that decision. They may now be tabulated and converted where appropriate to logarithms and the pro-forma made for the calculations except where the arithmetic is to be carried out on a computer.

Some statistics are expressed in different terms for different purposes. Manpower, for instance, is variously expressed in terms of man-hours, man-weeks, man-years, or simply as the average number of people employed. Choice will depend to some extent on convenience, and the number of significant figures needed; though it should be kept in mind that the several measures will tend to vary relatively to each other with the passage of time as the number of working hours a week or the length of holidays with pay are changed. If the output per man-hour in a factory increases by 10 per cent, and the average weekly hours worked fall at the same time from 50 to 40, the output per man-week will fall by 12 per cent. That is the kind of thing that happens in practice, so that the choice of measures is a matter of some importance. Generally, the choice will rest between man-hours and the average number employed in the period. The latter seems preferable for most purposes, as it eliminates the effect of the irrelevant disturbing factors consisting of changes in the working week and holidays with pay. The average number employed has the advantage over the man-year in so far as it can be used of any period without causing confusion.

SIMPLE LINEAR REGRESSION

A simple regression analysis is one that involves only two variables. The adjective *simple* as used in the literature refers to the problem and not to the method employed; and it is sometimes limited to linear problems. The simplest method is that of fitting a free-hand graph, or curve as it is usually called, to the scatter diagram of the co-ordinates of the two statistical series plotted on rectilinear graph paper. Strictly, the process of free-hand curve-fitting cannot often be applied in multiple regression, that is, in regression problems involving more than two variables.

Table 7.1 contains two series of statistics for variables X and Y; Y being regarded as the dependent variable. They are shown in the second and third columns of the table. The other columns should be ignored for the moment. A scatter diagram showing Y plotted against X is given in Fig. 7.1. As the dots of the co-ordinates appear to fall about a straight line, the relationship between X and Y can be said to be linear.

TABLE 7.1 Simple regression analysis: basic data

Period order			*Order of magnitude of X*					
Period	*X*	*Y*	*X*	*Total*	*Average*	*Y*	*Total*	*Average*
1	2	17	1			12		
2	5	30	2			17		
3	10	51	2			8		
4	6	40	4			30		
5	7	35	5	14	2·8	27	97	19·4
6	8	50	5			30		
7	1	12	6			40		
8	4	30	7			35		
9	5	27	8			50		
10	2	8	10	36	7·2	51	206	41·2
TOTAL	50	300		50	5·0		300	30
Average	5	30						

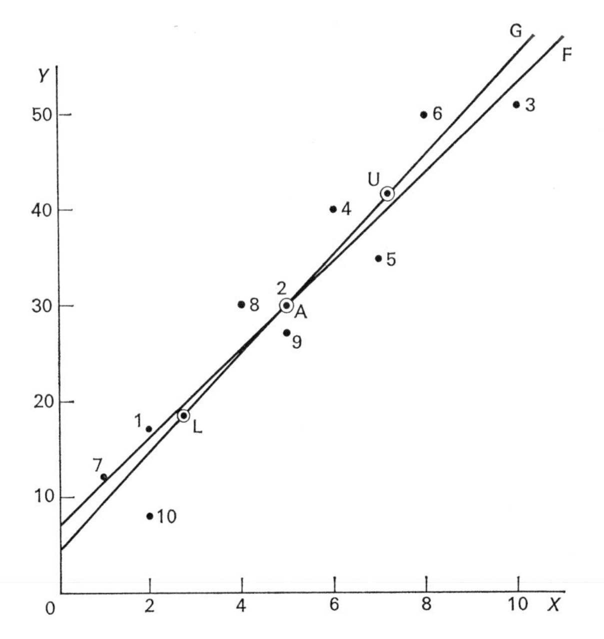

FIG. 7.1 Fitted curves: F by free-hand, G by group averages

For any straight line fitted to a scatter diagram, there is always one point through which the line should pass: it is the co-ordinate of the averages of the two series plotted. It is shown ringed and marked A in Fig. 7.1, the two averages being as shown in Table 7.1, i.e. $X=5$ and $Y=30$. The co-ordinate of the averages in a straight line relationship serves as a pivot for the line of best fit. It now remains to decide the slope of the line. One way is to use personal judgment, fitting the line so that it appears to pass midway through the scatter of dots at each end, leaving the pivot A to determine the middle scatter. Before the advent of transparent plastic rulers, it was common practice to use a piece of thin string pinned to the co-ordinate of the averages and rotated round it first one way then the other until the investigator had satisfied himself that he had achieved the best-fitting slope.

Two readings from the free-hand curve, preferably from the extremes to minimise the effect of errors in the readings, are now required to determine the linear equation that satisfies the statistical relationship between X and Y. In Fig. 7.1, the free-hand curve fitted by personal judgment through the co-ordinate of the averages, is marked F. For $X=0$, Y reads 7, which is the measure of the unattached constant. At the other end of the curve, we have $X=10$, $Y=53$. Of the latter, 7 is accounted for by the constant, leaving 46, which is attributable to the value of X. The value of the coefficient of X is therefore 4·6 (46÷10), and the complete equation to the free-hand curve is:

$$Y=4{\cdot}6\,X+7$$

which can easily be tested for accuracy by substituting one or two values within the limits of the diagram and comparing the resulting values of Y with the corresponding readings from the curve; e.g. for $X=6$, $Y=34{\cdot}6$, which tallies with the reading.

The method of group averages

A slightly more sophisticated means of determining slope than fitting a free-hand curve is the method of group averages, which necessitates the arrangement of the basic data into two or more groups by reference to the magnitude of one of the two variables. For this exercise, X has been chosen as the criterion, and two groups formed. The rearranged basic data are given in the last six columns of Table 7.1, which also shows the group totals and group averages. To continue graphically, the co-ordinate of each pair of group averages is then plotted on the scatter diagram as shown in Fig. 7.1, and a straight line curve is fitted through the pivotal co-ordinate marked A in the diagram, so that it passes at an equal distance above or below the group averages, or through them where the three averages fall into alignment as they do in Fig. 7.1.

Finally, in order to determine the equation to the curve, the same procedure as that employed above for the free-hand curve can be used. Reading from the graph we have for $X=0$, $Y=4{\cdot}5$, which is the value of the unattached constant, and for $X=8$, $Y=46$. Then the value of the coefficient of X is $(46-4{\cdot}5)/8$, which gives $5{\cdot}1$, the complete equation being

$$Y=5{\cdot}1\,X+4{\cdot}5$$

The degree of significance of the difference between this and the equation above derived from the free-hand curve depends upon the use to which the equation or either of the constants is put. If it is needed for estimating the value of Y for any given value of X, the difference is insignificant for values of X within the observed range. For values outside the observed range, the significance grows the farther away the values of X depart from the observed range.

As to the use of either the coefficient of X or the unattached constant for purposes extraneous to the equation, the difference may be highly significant. Unquestionably, of the two equations, the one determined from the line based on group averages is the more accurate; but there is no reason why both should not be used to give upper and lower limits of the entity being estimated, especially if it is felt that the free-hand curve provides as good a fit to the scatter of the dots as the curve fitted by group averages.

Testing for linearity of the model is automatic with the diagrammatic method. Where the co-ordinates of the group averages and of the overall average fall into alignment, as they do in Fig. 7.1, the result is positive and linearity is proved. Where they do not, and the deviation from a straight line is significant, a curvilinear form of model must be tested.

It will be observed that in Fig. 7.1, the dots of the co-ordinates are numbered. The numbers represent the periods as shown in the first column of Table 7.1, the supposition being that the basic statistics are time series. The purpose of the numbers is to provide a test for time trend in the Y series independent of the influence of X. Ten sets of observations are not enough for the purpose of the test, but Fig. 7.1 can be used for showing how the test is applied. If the dots for the later periods fall fairly consistently above or below those of the earlier periods, there is evidence that time plays a part in the magnitude of Y. It will be seen, for instance, that the dots numbered 1 and 10 fall on the same vertical, i.e. the value of X is the same for both. But dot 10 falls appreciably below dot 1, suggesting that the time trend of Y is downward. This tends to be confirmed by the dots numbered 2 and 9, which also happen to fall on the same vertical as each other; but here the distance between the two dots is much smaller, about a quarter of that between dots 1 and 10. The other dots have no time pattern at all. On the whole, it can be inferred that in spite of the pattern revealed by dots 1, 2, 9 and 10, there is no significant time trend.

It is sometimes possible to estimate the effect of time or other second independent variable on the dependent by the graphic method of inserting against each dot the value of this second independent variable, noting the vertical distances between the high values and the low values, dividing this distance by the difference between the two values in each case and averaging out. Suppose that in Fig. 7.1, there is a fairly consistent time pattern throughout on the lines of that suggested by dots 1, 2, 9 and 10. Using the readings from these four dots to demonstrate the method, we have $10-1=9$ periods and $9-2=7$ periods, the corresponding values of Y being $8-17=-9$ and $27-30=-3$. Supposing these provide enough evidence of a time trend, we then attempt to obtain an estimate of its measure. This is done by the method (or calculus) of finite differences, which forms the subject of the next section:

Periods No. (1)	*Change in Y* (2)	(2)÷(1) (3)
9	−9	−1·0
7	−3	−0·43
16	−12	−0·75

This means that from any one period to the next, Y decreases on average by 0·75, the appropriate equation (with X omitted) being:

$$Y=K-0{\cdot}75\ T$$

where K is the unattached constant, and T represents time. Needless to say, if we accepted this, our first estimate of the value of the coefficient of X would probably need amending. Problems like this in three or more variables are dealt with later in this chapter.

It is sometimes thought that negative values of variables occur only in mathematical contexts. The idea is a mistaken one: loss is negative profit, profit is negative loss; a decrease in turnover is a negative increase. In management statistics in common with statistics generally, the standard rules apply: 'two negatives make a positive' and so on. If two departments A and B together made a profit of £5,000 in 1970, and B made a loss of £1,000 what did A do?

	£
Profit of A and B	5,000
To determine A's profit deduct B's profit:	
Deduct	−1,000
∴ A's profit =	£6,000

The graphic method of regression analysis would be nothing if it did not permit the plotting of negative values. The standard complete form of diagram has four quarters known as quadrants. The vertical and horizontal axes provide the dividing lines. Figure 7.2 gives the complete form. The upper right-hand quadrant, often called the *north-east* (or NE) *quadrant*, is the one mostly used for statistical purposes. Figure 7.1, for instance, is the NE quadrant, the others being omitted as unrequired. If the best fitting straight line graph of the scatter in Fig. 7.1 were extended to the left for negative values of X, it would pass through the corner of the NW quadrant and enter the SW quadrant where the values of both X and Y are all negative.

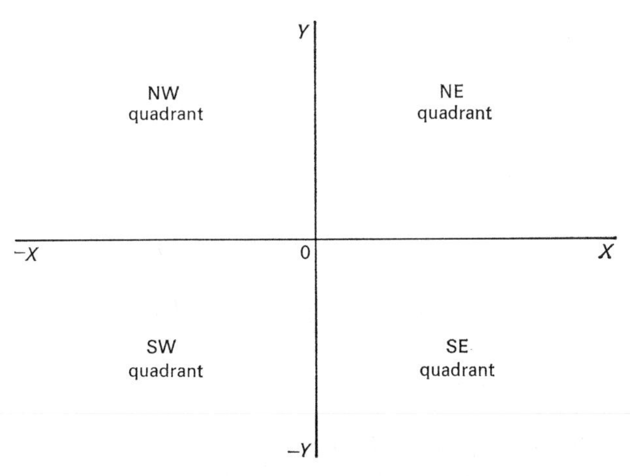

FIG. 7.2 The four quadrants

Graphic curtailment can often conveniently go beyond the omission of unrequired quadrants. Indeed, it is safe to say that in management statistics, it is often not merely convenient but also necessary to curtail an individual quadrant so that either one or both of the axes are omitted, and the scale of Y or of X or both are given on an internal vertical or horizontal. Curtailment is necessary where the range of a series is narrow relatively to the magnitude of the numbers it contains. This point was made in Chapter 2 in connection with graphic presentation; it applies equally to graphic analysis.

The method of finite differences

As a method of regression analysis, the calculus of finite differences does not depend upon graphing at all. It can serve as a means of determining the slope of a straight-line graph; it is a method in its own right. It is an adaptation of simultaneous equations to the analysis of statistical material. It can also be used in harness with the method of group averages. Consider the example of Table 7.1 for instance. Table 7.2

TABLE 7.2 The method of difference applied to group averages

	Average of X (1)	*Difference X* (2)	*Average of Y* (3)	*Difference Y* (4)	*Coefficient of X* (4)÷(2) (5)
Upper group	7·2		41·2		
Overall average	5·0	2·2	30·0	11·2	5·1
Lower group	2·8	2·2	18·2	11·2	5·1
		4·4		22·4	5·1

Solve for the unattached constant, b:

$$Y = 5 \cdot 1X + b$$

Substituting the overall averages for X and Y, we have

$$b = 30 - 5 \cdot 1 \times 5 \cdot 0$$
$$= 4 \cdot 5$$

shows how the method of finite differences applies simultaneous equations systematically in a suitable tabular form to facilitate the analysis of a multiplicity of data. The overall average is included partly for this reason, and partly to test the linear assumption of the underlying model. Finite differences can be applied to the individual sets of observations; but there does not seem to be a satisfactory way of averaging out the several estimates of the coefficient, other than by using the sums of the differences with signs ignored. Table 7.3 demonstrates the procedure applied to the example of Table 7.1.

TABLE 7.3 The method of differences applied to individual observations

X	ΔX	Y	ΔY	*Estimated coefficient of X* $\Delta Y/\Delta X$
2		17		
5	+3	30	+13	4·33
10	+5	51	+21	4·20
6	−4	40	−11	2·75
7	+1	35	−5	−5·00
8	+1	50	+15	15·00
1	−7	12	−38	5·43
4	+3	30	+18	6·00
5	+1	27	−3	−3·00
2	−3	8	−19	6·33
Sums of Δs ignoring signs	28		143	5·11

Note: Capital delta, Δ, is the sign of finite differentiation.

It will be seen that the sums of the differences give an estimated coefficient of X (5·11) which compares with the figure derived by the same method applied to group averages (5·10). The estimates derived from successive pairs of sets of observations vary appreciably, ranging as they do from −5·00 to +15·00. Although the two series of X and Y are highly correlated as Fig. 7.1 shows, the wide range of these estimates indicates the need for applying a smoothing device to the basic statistics before finite differences can be used with success. It also suggests that the figure of 5·11 based on the sums of the differences is not reliable—that its close approximation to what appears to be the correct answer is a mere coincidence. Indeed, a change in the order of the figures of the two series may result in an entirely different answer: the arrangement by order of magnitude of the X series shown in Table 7.1, for instance, gives $\sum \Delta X = 9$ and $\sum \Delta Y = 73$, so that the solution value of the coefficient of X is 8·11, which can scarcely be described as an approximation to the correct answer.

The method of least squares

Of the several methods of regression analysis, least squares is the most sophisticated, and since it is claimed that it conforms to the theory of error, it is the one generally preferred by statisticians. It has one drawback compared with other methods. It needs to be applied with care, and not treated as a kind of sausage machine, with the raw materials going in at one end and the finished product coming out at the other. Other methods give the analyst an opportunity of watching the regression behaviour of his raw materials in the processing, and also of testing the underlying model. Not so with least squares: nothing is seen of the regression behaviour of the basic statistics in the process of applying least squares; and there is no inbuilt means of testing the validity of the underlying model. The result is that there is a proneness amongst analytical statisticians to ignore the need for a model, to assume a linear relationship quite blindly, and to throw their raw materials into the least square machine without first considering their suitability for achieving the object in view.

None of this is to suggest that least squares is not a good method. Indeed it is by far the best provided it is used with reasonable care, and provided a valid model forms its basis and the basic statistics are properly prepared. A preliminary canter with one of the cruder methods such as the graphic or group averages is advisable: it helps in deriving and testing the appropriate model and in revealing the regression behaviour of the basic statistics.

In simple linear regression analysis, the model is:

$$Y = aX + b$$

When the best-fitting values of a and b have been determined, they can be applied to the observed values of X in order to obtain estimates for Y. Thus for the example of Table 7.1, we have on the group averages criterion of best-fit:

$$Y = 5{\cdot}1X + 4{\cdot}5$$

For period 1, the value of X is 2, so that the estimated value of Y is:

$$\begin{aligned} Y &= 5{\cdot}1 \times 2 + 4{\cdot}5 \\ &= 10{\cdot}2 + 4{\cdot}5 \\ &= 14{\cdot}7 \end{aligned}$$

compared with the actual value of 17. The difference is 2·3. The least squares criterion provides that the sum of the squares of the differences between the estimated and actual values of Y should be at a minimum, that is, the least of all the sums of squares of differences based on estimated values of Y calculated by any criterion. Hence the name, *least squares*.

It will be seen that the differences are the measure of the vertical distances between the dots of the co-ordinates and the fitted straight-line graph, so that it is the squares of these vertical distances that are minimised by the method of least squares. The method can also be used for minimising the squares of the horizontal distances. The former is known as the regression of Y on X, and the latter as the regression of X on Y. The regression of Y on X assumes that Y is the dependent variable, and X the independent, that is, that X influences Y, but that Y does not influence X, and that other factors also influence Y. The total effect on Y of other factors is referred to as *the error* and Y is said to carry its burden. If we could eliminate the effect on Y of these other factors, then, given accurate basic data, X and Y would be perfectly correlated.

However, if X happens to be the dependent variable, and Y the independent, then the regression of X on Y is appropriate. But if X and Y are interdependent, each influencing the other to some extent, as wage rates and the cost of living affect each other, then either least squares must be abandoned as the appropriate method, or some kind of average of the two regression equations must be struck. Attempts have been made to devise a least squares formula that would give what is called a mutual regression equation, presumably based on minimising the squares of the perpendicular distances of the dots from the straight-line graph; but so far as is known, without success. In any event, it is questionable whether it would be very useful: it would assume that the interdependence is balanced, that is, that Y depends upon X as much as X depends upon Y. In economics and industry, the interdependence of factors is common, but it rarely balances: it is probable, for instance, that wage rates influence the cost of living more than the cost of living influences wage rates. In striking an average, the weight to be given to each of the two regression equations must necessarily be a matter of subjective judgment, a process that rather defeats the design of least squares, which is to derive an equation by entirely objective criteria.

The formulae

For simple linear regression, the two least squares formulae are:

$$\text{Regression of } Y \text{ on } X: \sum(x^2)a = \sum(xy)$$

$$\therefore \quad a = \frac{\sum(xy)}{\sum(x^2)}$$

$$\text{Regression of } X \text{ on } Y: \sum(xy)a = \sum(y^2)$$

$$\therefore \quad a = \frac{\sum(y^2)}{\sum(xy)}$$

where x and y are, as before, the deviations from the means of the two series, and capital sigma, $\sum$, is the sign of summation. Table 7.4 shows how the two formulae are applied to the example of Table 7.1.

TABLE 7.4 Simple linear regression by least squares

	Basic series				
	X	Y	X^2	Y^2	XY
	2	17	4	289	34
	5	30	25	900	150
	10	51	100	2,601	510
	6	40	36	1,600	240
	7	35	49	1,225	245
	8	50	64	2,500	400
	1	12	1	1,444	12
	4	30	16	900	120
	5	27	25	729	135
	2	8	4	64	16
Totals	50	300	324	10,952	1,862
Averages	5	30	—	—	—
Correction sums			250	9,000	1,500
Sums of squares etc. of deviations			74	1,952	362
			$=\sum(x^2)$	$=\sum(y^2)$	$=\sum(xy)$

Regression of Y on X: $\frac{362}{74} = 4{\cdot}91$

Regression of X on Y: $\frac{1{,}952}{362} = 5{\cdot}39$

Solve for b in the way demonstrated above, i.e.:

$$b = Y - aX$$

Substituting the averages of Y and X, we have:

$$\text{For } Y \text{ on } X: b = 30 - 5 \times 4{\cdot}91 = 5{\cdot}45$$

$$\text{For } X \text{ on } Y: b = 30 - 5 \times 5{\cdot}39 = 3{\cdot}05$$

Like the method of finite differences, least squares is founded on simultaneous equations, and both regressions give precisely the same answer as each other and as simultaneous equations and finite differences when applied to the minimum number of sets of observations required to give a solution. For problems in two variables, two sets of observations are required. Applying least squares to the first two sets in Table 7.1, we have:

	X	Y	X^2	Y^2	XY
	2	17	4	289	34
	5	30	25	900	150
Totals	7	47	29	1,189	184
Correction			24·5	1,104·5	164·5
Sums etc.			4·5	84·5	19·5

$$\text{Regression of } Y \text{ on } X: \frac{19{\cdot}5}{4{\cdot}5} = 4{\cdot}33$$

$$\text{Regression of } X \text{ on } Y: \frac{84{\cdot}5}{19{\cdot}5} = 4{\cdot}33$$

It will be seen from Table 7.3 that this agrees with the solution coefficient of X reached by finite differences. By simultaneous equations, we have:

$$17 = 2a + b \qquad \ldots (a)$$

$$30 = 5a + b \qquad \ldots (b)$$

Eliminate b by deducting (a) from (b):

$$13 = 3a$$

$$\therefore \quad a = 4{\cdot}33$$

SIMPLE CURVILINEAR REGRESSION

As previously implied the processes and procedure described and demonstrated above in respect of linear regression apply to curvilinear regression. With models of the logarithmic type, the whole of the basic data are first converted to logarithms. The application of regression analysis to models of quadratic or similar type will be examined later.

Time trends

Time trends provide something of a special case. Oddly enough, many textbooks on statistics demonstrate the method of least squares by applying it to a time trend, but always on the assumption that the trend describes a straight line on rectilinear graph paper. There is no difficulty at all in applying any method of regression analysis to a linear trend like that. A linear trend is merely an arithmetic progression, and the series inclines upward or downward on average by an absolute amount in each period. The difficulty arises where the trend describes a geometric progression, since to use logarithms for time as well as for the series would give a misleading solution. Time must remain in its natural form for the purpose. Suppose the base term in the time series is 1,000, and the common ratio is 1·05, then the first term is $1{,}000 \times 1{\cdot}05$, and second is $1{,}000 \times 1{\cdot}05 \times 1{\cdot}05$, and nth term is $1{,}000 \times 1{\cdot}05^n$. Let b represent the constant, which is the base term; r, the common ratio; and Y the dependent variable. Then we have:

$$Y = br^n \qquad \ldots \text{(i)}$$

or in linear form:

$$\log Y = \log b + n \log r \qquad \ldots \text{(ii)}$$

and:

$$\log r = \frac{1}{n} (\log Y - \log b) \qquad \ldots \text{(iii)}$$

In equation (ii), $\log r$ is the coefficient of n, the independent variable representing time. Since $\log b$ is an unattached constant, equation (ii) can be regarded as a valid linear model to which it is possible to apply any method of simple regression analysis including least squares.[1]

Equation (iii) provides a method of ascertaining the percentage rate of increase or decrease per period. It is often said that the real gross national product increases at an average annual rate of 3 per cent. The truth of this can be tested by applying equation (iii) to the GNP at constant prices for the years, say, 1959 and 1969. The GNP at

[1] See equations (i)–(iii) of Chapter 3 for the actuarial form of these equations; and Chapter 13 (p. 239) for a demonstration of the practical value of equation (ii).

factor cost in terms of 1963 prices amounted to £23,759 million in 1959 and to £31,695 million in 1969. We therefore have:

$$\log r = \frac{1}{10}(4{\cdot}5010 - 4{\cdot}3758)$$

$$= 0{\cdot}01252$$

$$\therefore\ r = 1{\cdot}029$$

The annual rate of increase is therefore 0·029, the percentage rate being 2·9, so that the oft-quoted figure of 3 per cent is a good rough approximation for general use. For analytical and forecasting purposes, however, even 2·9 per cent can scarcely be accepted. For such purposes, the least squares trend (the regression of Y on n) should be fitted to the GNP statistics of the last ten years or so.[1]

Quadratics and similar equations

Although theoretically quadratics and higher-degree equations have only a small part to play in the field of economics and industrial management, there may be occasions in practice when an empirically derived model takes on the shape of such an equation. Equations of this type have two peculiar features for regression analytical purposes. Although they are concerned with only two variables and show a curvilinear relationship, they are treated for regression purposes as linear problems in more than two variables. Consider the equation:

$$Y = K + aX + bX^2 + cX^3$$

where K is the unattached constant and a, b, and c are the coefficients of X, X^2 and X^3. It can be rewritten:

$$Y = K + aX + bU + xV$$

where $U = X^2$ and $V = X^3$. This equation has all the appearances, and indeed is, a linear equation in four variables. The solution calls for multiple correlation analysis, which forms a subject of later sections of this chapter.

However, this is a convenient place to consider the application of finite differences to determining quadratic and similar relationships.

[1] GNP statistics for UK are published annually in the *National Income* blue book (HMSO). Figures at 1963 prices appear in Table 14 of the 1970 issue.

TABLE 7.6 Data plotted in Fig. 7.3

X	Y
1	7
3	30
5	70
4	40
2	14
7	110
5	63
6	90
9	200
8	145
10	220

As stated above, the method of differences calls for smoothed data, that is, data that have been rendered smooth either by group or moving averages or by fitting a curve to the scatter of the co-ordinates. Table 7.6 contains an example of two series which need to be smoothed and tested

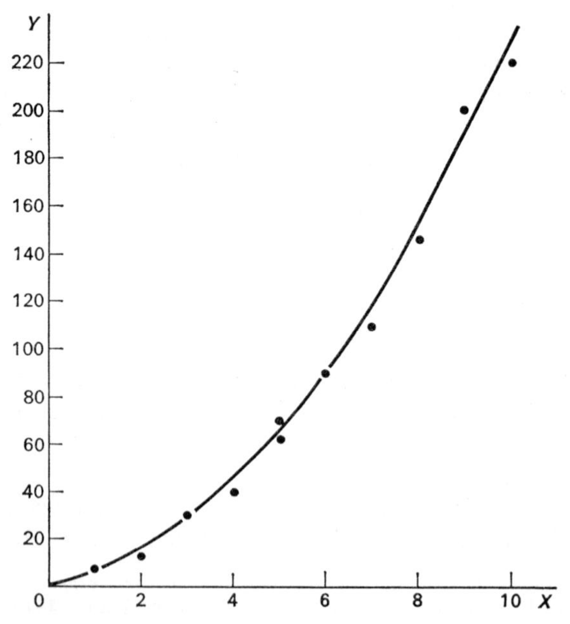

FIG. 7.3 Smooth curve fitted to dot diagram

for type of model. It is considered that Y is the dependent variable and that X is the major factor influencing the value of Y. Figure 7.3 shows the scatter of the dots of Y plotted against X, and a smooth free-hand curve fitted to the dots.

The next step is to take a number of readings from the curve, and analyse them in the way demonstrated in Table 7.7.

TABLE 7.7 Analysis of curve in Fig. 7.3 by finite differences

Line	(1)	(2)	(3)	(4)	(5)	(6)	(7)	(8)	(9)
Readings from graph									
(*a*)	X	1	2	3	4	5	6	8	10
(*b*)	Y	6	15	28	45	66	91	153	231
Differences—First series									
(*c*)	X'		1	1	1	1	1	2	2
(*d*)	Y'		9	13	17	21	25	62	78
(*e*)	Y'/X'		9	13	17	21	25	31	39
Differences—Second series									
(*f*)	X''			2	2	2	2	3	4
(*g*)	Y''			4	4	4	4	6	8
(*h*)	Y''/X''			2	2	2	2	2	2

Notes:
Line (*c*) is derived from line (*a*): $2-1, 3-2 \ldots 10-8$;
Line (*d*) from line (*b*): $15-6=9 \ldots 231-153=78$;
Line (*f*): from line (*a*): $3-1, 4-2 \ldots 8-5, 10-6$;
Line (*g*): from line (*e*): $13-9=4 \ldots 39-31=8$
The X differences in each series are always derived from the original readings, whereas the Y differences are always derived from the Y/X figures of the preceding series. In the table, the trendlessness of line (*h*) which gives the Y/X figures of the second series indicates that the graph of Fig. 7.3 is a quadratic, and the series of figures in line (*h*) shows that the coefficient of X^2 is 2.

Table 7.7 serves the twofold purpose of giving the degree of the equation and the value of the coefficient of X^2. In a case where the Y/X figures of the second series do not prove to be trendless, the process needs to be continued into a third series and so on, until a trendless series is achieved, or the analysis abandoned on the grounds that the relationship between the two basic series is not of the type. Each successive series of Y/X should record a reduction of the upward or downward trend. If it records an increase in any series, then the relationship is not of the type, and the analysis is best abandoned.

To complete the equation to the graph, the influence of $2X^2$ needs first to be eliminated from the series of Y readings:

Readings of Y	6	15	28	45	66	91	153	231
$2X^2$	2	8	18	32	50	72	128	200
Residue of $Y = Z$	4	7	10	13	16	19	25	31
Differences Z'		3	3	3	3	3	6	6
Z'/X'		3	3	3	3	3	3	3

Therefore the coefficient of X is 3. If $3X$ is then deducted from the Z series, the final residue is the unattached constant, which a glance at these figures shows to be 1, the complete equation to the graph being:

$$Y=2X^2+3X+1$$

Where the final Y/X series is a constant as it is in Table 7.7, the readings from the graph give an ideal basis for the analysis, and there is no need to take further readings.

It goes without saying that the results of an analysis of this kind are highly sensitive to misreadings from the graph. The diagram needs to be large to facilitate accurate measurement. The original chart used for plotting, graphing and taking readings for the example of Table 7.6 was 10 in. by 10 in., which proved to be large enough for the purpose. Another important point is that taking a large number of readings from the graph for analytical purposes is usually worth while, especially where the graph has a double bend in it indicating that it is of the third or higher degree; for each successive series reduces the number of effective pairs of values by one. A fifth degree equation would call for five successive series, and would reduce eight original sets to three which is scarcely enough for indicating trend.

It should be kept in mind that when averaging has to be resorted to for smoothing, the averages, where the relationship between the two factors is curvilinear, are like the rear wheels of a car: they tend to take a short cut round bends and corners. One way to minimise this tendency is to use as many observations as possible for averaging on bends in the graph thus keeping the sector of graph covered by an average within reasonable limits. The trouble is that averaging assumes linearity over the sector of the graph covered. Where the graph tends to straighten out, the sector covered by an average can be much longer. This applies only to the process of averaging for quadratics and similar relationships when the natural numbers have to be used. For logarithmic relationships, there is no such difficulty provided the logarithms of the two series are used for averaging.

MULTIPLE REGRESSION ANALYSIS

The methods of multiple regression analysis are basically much the same as those used in simple regression analysis. Group averages necessitate the arrangement of the basic material into classes by reference to order of magnitude, the object being to examine the influence of each independent variable in turn while the other or others are held constant. For this reason, the term *group averages* is dropped, and the method becomes *cross-classification.* Finite differences can be applied to the group averages of cross-classification, just as they can in simple regressions, to arrive at the equation of best fit. There is one important additional auxiliary method of use in multiple regression: it is called successive elimination. It is important largely because its application helps the analyst to observe the regression behaviour of his basic material. Finally, there is least squares, which reigns supreme amongst the methods of regression analysis in the multiple field as in the simple one.

Cross classification

Appendix 1 of my book on *Planning Profit Strategies* contains in Table A1.1 a linear analysis of driver man-hours engaged on the collection and delivery of packages in a town or urban locality by motor van. The analysis is carried out by cross-classification, and it tends to confirm empirically the conclusion previously arrived at theoretically that the relationship between the dependent variable, i.e. the total number of weekly man-hours spent on each vehicle on the one hand, and the independent variables on the other, was entirely linear. The independent variables are mileage and the number of effective calls made, both per vehicle per week. The appropriate model is:

$$H = aM + bE + K$$

where H is the total number of man-hours per vehicle per week; M and E are the corresponding mileage and effective calls made respectively; and K is a constant representing the man-hours per vehicle per week spent on such things as preparation, loading and unloading at the depot and meals. The example provides 35 sets of observations cross-classified into group averages with the number of calls shown across the table and the mileage downward.

There is no need to display the detail. All that is necessary for our purpose here are the averages, which are given in Table 7.8. The table indicates the arrangement of the detailed sets of observations.

It will be seen that the cross-classification has not entirely eliminated changes in the number of calls read downward or the vehicle miles read across as it was hoped. The averages indicate a degree of negative correlation between the two variables, and this largely explains the failure to render either factor constant. Correlation between independent variables is bound to make cross-classification difficult. The most nearly constant set of averages is that of vehicle-miles in line B: they read 114·5, 113·2 and 112·0. Suppose they are constant, then the procedure would be first to solve for b by using the figures for E and H in line B:

E	ΔE	H	ΔH	$b = \Delta H/\Delta E$
42·5		64·3		
68·0	25·5	65·0	0·7	0·0274
125·0	37·0	67·3	2·3	0·0622
	62·5		3·0	0·0480

As will be seen later, the average of 0·048 is a fair approximation to the value of b.

Finite differences

However, the averages derived by cross-classification can be used as smoothed figures for analysis by finite differences. Table 7.9 sets out the complete solution. As the two extreme sets of averages, those in main cells A1 and C3 in Table 7.8, are based on only two sets of observations each, they are omitted from Table 7.9 as unrepresentative. Line (g) consists of successive ratios, not differences, e.g. $10{\cdot}5/25{\cdot}5 = 0{\cdot}412$. The several estimated values of a, the coefficient of M, are remarkably consistent; and there seems no good reason why their arithmetic mean of 0·5 should not be adopted as reasonably accurate. The exercise is a systematic way of applying simultaneous equations to an array of statistics: the first differences eliminate K; and the second differences,

E, leaving the residual values of H in line (k) as entirely attributable to M in line (l).

TABLE 7.8 Driver man-hours per vehicle per mile

No. of calls	*1. Up to 50*			*2. 50–100*			*3. Over 100*		
Vehicle-miles	*Miles*	*Calls*	*Hours*	*Miles*	*Calls*	*Hours*	*Miles*	*Calls*	*Hours*
A up to 100	85·0	45·0	51·0	77·5	75·0	48·0	75·0	135·0	49·3
B 100–120	114·5	42·5	64·3	113·2	68·0	65·0	112·0	125·0	67·3
C Over 120	123·6	32·0	68·4	125·0	70·0	71·2	122·5	115·0	72·0

TABLE 7.9 Finite differences applied to the example of Table 7.8

	(1)	(2)	(3)	(4)	(5)	(6)	(7)	(8)
Averages as in Table 7.8:								(Overall)
(a) H	68·4	64·3	65·0	71·2	48·0	67·3	49·3	—
(b) M	123·6	114·5	113·2	125·0	77·5	112·0	75·0	—
(c) E	32·0	42·5	68·0	70·0	75·0	125·0	135·0	—
First differences:								
(d) H'		−4·1	0·7	6·2	−23·2	19·3	−18·0	−19·1
(e) M'		−9·1	−1·3	11·8	−47·5	34·5	−37·0	−48·6
(f) E'		10·5	25·5	2·0	5·0	50·0	10·0	103·0
(g) E'' (successive ratios)			0·412	12·25	0·40	0·10	5·0	—
(h) $H'E''$			0·29	75·95	−116·00	1·93	−90·00	—
(j) $M'E''$			−0·54	144·55	−237·50	3·45	−185·00	—
Second differences:								
(k) H''				75·66	−191·95	117·93	−91·93	—
(l) M''				145·09	−382·05	240·95	−188·45	—
(m) H''/M'' (estimated value of a)				0·521	0·502	0·490	0·488	—

Solve for b by using 0·5 for a over the full range of E given in col (8):

(n) aM'	−24·3
(p) $H'-aM'$	5·2
(q) $(H'-aM')\div E'=b$	0·0505

Solve for K by using $a=0\cdot 50$ and $b=0\cdot 051$ applied to the overall averages, viz. $H=62\cdot 4$, $M=106\cdot 9$ and $E=76\cdot 6$:

$$K=62\cdot 4-0\cdot 50\times 106\cdot 9-0\cdot 051\times 76\cdot 6$$
$$=5\cdot 0$$

Complete equation:

$$H=0\cdot 50M+0\cdot 046E+5$$

Successive elimination

Successive elimination is a method in its own right; but its principal value lies in its use as a means of correcting the coefficients estimated by cross classification. It is a method of trial and error and is best applied not to the figures of the basic series but to cross-group averages. Table 7.10 demonstrates the method applied to the same seven cross-group averages used in Table 7.9. First the averages are arranged in order of magnitude of one of the two independent variables and the first differences extracted to eliminate the unattached constant. Then the independent variable which determined the order of the data is eliminated from the first differences by using a coefficient which may be pure guesswork or may be estimated from a cross section of the cross-classification matrix as shown above in the small table on p. 118. The next stage is to calculate from the residual values of the dependent variable and the first differences of the other independent, the resulting estimated values of the coefficient of this other independent.

TABLE 7.10 Successive elimination applied to three variables

	(1)	(2)	(3)	(4)	(5)	(6)	(7)
First differences from Table 7·9:							
(*a*) H'	−4·1	0·7	6·2	−23·2	19·3	−18·0	−19·1
(*b*) M'	−9·1	−1·3	11·8	−47·5	34·5	−37·0	−48·6
(*c*) E'	10·5	25·5	2·0	5·0	50·0	10·0	103·0
Eliminate E (try $b = 0{\cdot}10$)							
(*d*) $0{\cdot}1\ OE'$	1·05	2·55	0·20	0·50	5·00	1·00	10·30
(*e*) $H'-0{\cdot}1OE'=H''$	−5·15	−1·85	6·00	−23·70	14·30	−19·00	29·40
(*f*) $H''/M'=a=$	0·566	1·423	0·509	0·500	0·415	0·514	0·605
Variation too great—try $b=0{\cdot}08$							
(*g*) $0{\cdot}08E'$	0·84	2·04	0·16	0·40	4·00	0·80	8·24
(*h*) $H'-0{\cdot}08E'-H'$	−4·94	−1·34	6·04	−23·60	15·30	−18·80	−27·34
(*j*) $H''/M'=a$	0·543	1·03	0·512	0·497	0·444	0·508	0·563
Variation reduced but still too great—try $b=0{\cdot}05$							
(*k*) $0{\cdot}05E'$	0·53	1·28	0·10	0·25	2·50	0·50	5·15
(*l*) $H'-0{\cdot}05E'=H'$	−4·63	−0·58	6·10	−23·45	16·80	18·50	24·25
(*m*) $H''/M'=a=$	0·509	0·447	0·517	0·494	0·487	0·500	0·499

This is satisfactory: the value of b can be taken as 0·05 and that of a as 0·50. The value of the unattached constant is calculated in the way demonstrated at the foot of Table 7.9.

Finally, an examination is made by inspection and comparison of these values of the coefficient of the other independent. If they are very erratic or have a significant upward or downward trend, the results are unacceptable, and another estimated value of the coefficient should be tried. If the resulting values of the other coefficient are less erratic or have a less pronounced trend, the second estimate lies on the correct side of the original estimate. In the example of Table 7.10, the first estimate of the coefficient of E is 0·10, which is known to exceed the probable correct value, merely to show how erratic the estimates of the coefficient of M can be. The second estimate, 0·08, results in values that are much less erratic, which goes to show that the second estimate is nearer the truth than the first; but it still looks as though it could be improved upon. Finally, 0·05 is tried, and this gives a series of values which are only slightly erratic. Figures of 0·049 and 0·051 could be tried, but it is questionable whether any further improvement would result. The figures in column (7) are the overall difference between the highest and lowest of the sets of averages. They also equal the sums of the first differences in columns (1) to (6). They are more likely to give an accurate solution than any of the individual columns, and the final result provides a standard of comparison for the results obtained in those columns.

Least squares

What, it may be asked, is the good of cross-classification, finite differences and the rest when there is least squares? The twofold answer is given above, but it is worth repeating for emphasis. First, analysis by one of these less sophisticated methods assists in determining the best-fitting type of model; and secondly, it gives the analyst an insight into the regression behaviour of his material. It is a mistake to suppose that a statistician who is familiar with his material also necessarily understands its regression behaviour. A knowledge of the regression behaviour of statistical data can be acquired only by applying to them one of the less sophisticated regression methods like those described and demonstrated above. Not that other kinds of knowledge of the particular sphere of statistics is useless. Far from it: statistical analyses of all kinds call for a knowledge of the precise definitions of the basic material at least. There is another point: mathematical statisticians have developed formulae for calculating margins of error for given probabilities for regression analysts; and much as these efforts are to be applauded and the results appreciated, there are practical statisticians who place equal or greater store on acquiring an insight into the regression behaviour of the basic material.

In the closing stages of a multiple regression analysis by least squares, a number of equations for simultaneous solution are produced. These equations are in effect the basic least squares formulae, which are very easy to remember. The formula for any number of variables is arrived at by using the following pro-forma:

$$
\begin{array}{cccccccc}
 & x_2 & x_3 & \ldots & x_n & & x_1 \\
x_2 & x_2^2 & x_2x_3 & \ldots & x_2x_n & = & x_1x_2 \\
x_3 & x_2x_3 & x_3^2 & \ldots & x_3x_n & = & x_1x_3 \\
 & & & \ldots & & & \\
 & & & \ldots & & & \\
x_n & x_2x_n & x_3x_n & \ldots & x_n^2 & = & x_1x_n \\
x_1 & x_1x_2 & x_1x_3 & \ldots & x_1x_n & = & x_1^2
\end{array}
$$

where x_1, $x_2 \ldots x_n$ are the deviations from the mean of the variables X_1, $X_2 \ldots X_n$, X_1 being the adopted dependent variable. This forms the basis of the equations. They are completed by adding the summation and plus signs and the symbols representing the coefficients. The set of equations of a three-variable analysis, for instance, is:

$$\text{A} \quad \Sigma(x_2^2)a_2 + \Sigma(x_2x_3)a_2 = \Sigma(x_1x_2)$$

$$\text{B} \quad \Sigma(x_2x_3)a_2 + \Sigma(x_3^2)a_3 = \Sigma(x_1x_3)$$

$$\text{C} \quad \Sigma(x_1x_2)a_2 + \Sigma(x_1x_3)a_3 = \Sigma(x_1^2)$$

in which x_1 represents the deviations from the mean of the variable X_1 which is adopted as the dependent variable for the purposes of the analysis. Since the use of deviations removes the unattached constant from the equations, only two equations are required to arrive at the solution. Where X_1 is known to be the true dependent, equations A and B are used, and C abandoned. Where X_2 is known to be the dependent, then the appropriate equations to use are B and C; and where X_3 is the dependent, the equations to use are A and C. It is questionable, however, whether anyone would want to adopt as the dependent variable a variable which is known to be independent. The linear model is:

$$X_1 = X_2a_2 + X_3a_3 + K$$

where the basic series are expressed in terms of natural numbers for linear relationships and quadratics, or in terms of logarithms for exponential relationships. If X_2 happened to be the known dependent variable, then conversion to the equation which expresses X_2 as a function of X_1 and X_3, that is, to the mathematically logical equation in the circumstances, would be as follows:

$$X_2a_2 = X_1 - X_3a_3 - K$$

$$\therefore X_2 = X_1\left(\frac{1}{a_2}\right) - X_3\left(\frac{a_3}{a_2}\right) - \left(\frac{1}{a_2}\right)K$$

It follows that once the values of a_2 and a_3 have been estimated, conversion is a matter of simple arithmetic.

However, all three of the least squares equations A, B and C above for a problem in three variables, come into their own where there is no clear-cut dependent variable as often happens in the field of business management. Then for a three-variables problem there are three solution values of each parameter, one from solving equations A and B, one from B and C, and one from A and C. For a problem in n variables, there are n solutions from n equations, of which there is always one too many for a single solution.

When the several solutions have been calculated, the next step is to strike an average for each parameter, and this will depend upon a subjective judgment of the weights to be applied to each, which in turn will depend upon a judgment equally subjective of the force exerted by each variable on each of the others, and the degree of independence enjoyed by each. Statistical theory assumes there is a clear cut and theoretically ascertainable dependent variable. The result is that much of the literature devoted to statistical theory is unrealistic from the management statistician's point of view.

It is proposed to devote the rest of this chapter to a regression analysis by least squares of the market in gilt-edged securities. A real-life exercise like this will give an idea of the kind of thinking that must necessarily underlie a satisfactory regression analysis. There is a twofold reason for choosing the bond market for the purpose. First, the published statistical material is ample. And secondly, although the bond market is not of major interest to management statisticians, it is of general interest to the run of management accountants, and of special interest to those statisticians concerned with the investment of funds such as the funds of insurance companies and pension schemes.

A REGRESSION ANALYSIS OF THE GILT-EDGED MARKET

By far the most important consideration that active and potential investors in gilt-edged securities take into account is the trend and level of the market prices. Of the several factors influencing the market, price is therefore the most convenient to adopt as the dependent variable. It is also probably the most nearly dependent, though there may be some interaction between the level of bond prices and at least one other factor, viz.: the trend and level of equity prices, which influence rather than are influenced by, bond market prices.

Government bonds are issued for varying periods ranging from about two years to upwards of 40 years. They are identified by reference to name, the rate of interest payable on the nominal value and the year or years of redemption. Treasury stock, 6¾ per cent, 1974 was issued in the latter half of 1970, the total life to redemption being about 4 years; in contrast, Treasury stock, 5½ per cent, 2008–12 was issued in 1964, the total life to redemption being 44–48 years. A given bond may not be issued on a single day. *The Financial Times* Share Information Service quotes two bonds that are *on tap*; they are Treasury, 6½ per cent, 1976 'A' and Treasury, 9 per cent, 1994 'A'. The Government broker has supplies of each available for sale, that is, the bonds are on tap, at their going market price.

Most bonds are issued at par; the exceptions usually arising when the market takes a turn for better or worse after the first date of issue. Some Treasury stock 5½ per cent 2008–12 referred to above was issued at 95 per cent of the nominal price when the market took a turn for the worse in 1964. The issue price and time of issue are of no concern to the market. What matters are the current market price and the date of redemption, the latest date being the criterion.

There are two internal factors affecting market price, one is the redemption date and the other is the nominal rate of interest often referred to as the *coupon*. On the whole, the market prices of bonds at any given time are governed by the laws of supply and demand so that for any given redemption date, the rate of yield based on the market price is much the same for different stocks, those with a low coupon having a lower market price than those with a high coupon.

These remarks apply to stocks that have a definite redemption date. There are some undated stocks, which include Conversion stock, 3½ per cent, 1961 *or after*.

There are two rates of yield based on the market price: one is the annual rate of interest gross of tax paid on the bonds by the Government, and the other is the gross redemption yield which takes into account the capital appreciation that takes place during the remaining life of a bond, or the capital depreciation if the market price happens to exceed the par value. All Government bonds are redeemed at their par value on the day of redemption, so that if the market price of a £100 bond stands at £75 today, it will appreciate by £25 during its remaining life. This figure of £25 is spread over the remaining life by a method of discounting and the annual amount is added to the interest yield to give the gross redemption yield. According to *The Financial Times* of 21 January 1971, the closing market price of Electrical stock, 3 per cent, 1974–7 was £82·25 per £100; the interest yield was 3·7 per cent and the gross redemption yield 6·8 per cent.

The available statistics

The available statistics dictate the factors to be taken into account, and also to a small extent the type of the model. In this exercise, there is no shortage of material. There are two internal factors to the market price, viz.: the coupon and the remaining life. In *The Financial Times* Share Information Service, a selection of over 40 Government securities including six undated ones is given daily, and it shows the coupon, the year or years of redemption, the price at the previous working day's close, the interest yield, and for the dated stocks, the gross redemption yield. The data are arranged in chronological order of redemption year beginning with the *shorts* or *short-dated* stocks, i.e. those with short remaining lives, and terminating with the long-dated and undated stocks.

A glance down the series of gross redemption yields shows a fairly consistent increase in the figures as the remaining life increases. On 20 January 1971 those with a life of up to five years had a redemption yield averaging 7 per cent, whereas those with a life of upwards of 15 years including undated stocks had a yield averaging nearly 9½ per cent. It follows that for any given coupon, prices are much higher at the short dated end of the market than they are at the long dated end.

An odd feature about the redemption yields is that for any given life they tend to be lower for low coupons than for high coupons. The lowest surviving coupon amongst the dated stocks is 3 per cent. Seven of these are given in the F.T. Share Service. In addition there are some at 3½ per cent coupon and a few at 4 per cent. All these stocks were issued at a time when interest rates were very low, that is, during the years 1932 to 1951, when both interest yields and redemption yields of long-dated and undated stocks ranged from 2½ to 4 per cent. (For short-dated stocks, they ranged from 2 to 3 per cent.) In those days, gilt-edged securities were held in high esteem as trustee stocks; and many inactive investors bought them for their interest yield, often on the advice of bank managers and stockbrokers. Inactive investors generally are not concerned at all with the market price of their holdings; so long as they receive their interest warrants, all is well. Consequently, such bonds tend to remain in short supply, so that their prices retain and continue to retain a measure of buoyancy relative to stocks that were issued since 1951, when the bond market began turning downward and became less stable and more speculative.

Nevertheless, the price of a stock with a coupon of, say, 5 per cent stands at a higher level than that of a stock with a coupon of 3 or 4 per cent and the same life. Three issues with remaining lives of 6–7 years demonstrate the phenomenon in a nutshell:

Stock	*Coupon*	*Date*	*Price*	*G.R. yield %*
Electrical	3 per cent	1974–7	82·25	6·8
Transport	4 per cent	1972–7	83·25	7·2
Exchequer	5 per cent	1976–8	85·10	7·9

These are typical of the market as a whole. The phenomenon has a disturbing influence which cannot easily be eradicated.[1] It means that the degree of statistical correlation between the coupon and the market price may not be so high as one would expect it to be. Indeed, as we shall see, owing partly to the phenomenon and partly to a high degree of correlation between the coupon and the life, the coefficient of correlation between coupon and price turns out to be remarkably low.

It will be appreciated that the coupon and the life are factors which account for variations in market price only at any given time. They do not account for changes in the general level of the market, such as took place between 1951 and November 1970, when the interest yield of 2½ per cent Consols rose from an average of 3·78 to 9·91 per cent.

It seems that the two main factors responsible for changes in the general level of bond market prices are the state of the equities market mentioned above, and changes in the purchasing power of money. As equities rise gilt-edged fall and *vice-versa*. Financial commentators compare the dividend yield of equities and the interest yield of undated

[1] One way is to leave low coupons out of account. A footnote to price index numbers of bonds in *Financial Statistics* reads, 'From January 1963 stocks with coupons below 4 per cent were omitted from the calculations and from January 1966 the limit was 5 per cent.'

stocks. When the former exceeds the latter, the excess is referred to as the *yield gap*; when it falls short of it, the shortfall is referred to as the *reverse yield gap*. A yield gap persisted from 1926 to 1959, since when a reverse yield gap has persisted. Industrial profits are a minor factor in this phenomenon. By far the most important factors are a rise in equity prices and a fall in bond prices. In spite of the fall in equity prices since January 1969, the reverse yield gap was still very wide on 20 January 1971: the average dividend yield, according to the F.T.–Actuaries share index numbers, stood at $4\frac{1}{2}$ per cent; and the interest yield of $2\frac{1}{2}$ per cent Consols, at $9\frac{1}{2}$ per cent.

The phrase *changes in the purchasing power of money* is a euphemism for deflation and inflation. In the years 1920 to 1934, it meant deflation; in the years since 1940, it has meant inflation. It is not changes in retail prices and therefore in the purchasing power of money that affect the bond market and interest rates generally so much as changes in the rate of change. Bond prices remained high and interest rates low up to 1951. In that year retail prices rose by an unprecedented 10 per cent, financial commentators began pointing out the folly of holding fixed-interest bearing securities with a yield of about 3 per cent, scarcely enough to make good a third of the loss in the real value of the capital invested in them let alone provide an income, and many investors in the gilt-edged market saw the writing on the wall for the first time, began selling their holdings, and with the proceeds buying equities, which, argued the commentators, generally retain their real capital value in times of inflation owing to the resulting rise in profits.

Equity prices continued their upward movement, and bond prices their downward movement. Then in January 1969, there was a sudden change; equity prices, it seems, had overreached themselves and began falling, and they tended to drag down bond prices with them until the autumn of 1969, when bond prices began to recover. However, the recovery was not destined to last for more than a few months; for they took a strong downward turn in April 1970, largely, it seems, in consequence of an increase in the rate of inflation. In theory, other things being equal, bond prices adjust themselves to give a yield which covers the current rate of inflation and in addition provides a fair return on investment. If retail prices are rising at, say, 5 per cent a year, and a fair return on capital is, say, 4 per cent, then bond prices tend to settle down at a level that gives a total yield of 9 per cent. Other things being equal, the figure of 9 per cent will persist until the rate of increase in retail prices changes. Statistically then, the appropriate series representing the factor of the purchasing power of money or its inverse, retail prices, must be expressed in terms of the rate of change in the recent past.

The model

Both theoretically and empirically, the model is of the logarithmic type. The factors that give rise to a change in the level of bond prices cannot be additive, they must enter the equation as multipliers. The relationship between price and these factors is therefore not linear but logarithmic. Graphing and group averages using bond prices and each of the more important factors in turn empirically confirm the theoretical conclusion.

Since the available data are of two kinds, one kind relating to different bonds at the same time, and the other, in effect, relating to any given bond at different times, and since the former holds extraneous factors constant, and the latter applied to an undated bond holds internal factors constant, the model can conveniently take the form of two equations, namely, in linear form:

$$\log B = a \log E + b \log R + \log K_1 \qquad \ldots \text{(i)}$$

$$\log P = c \log C + d \log T + \log K_2 \qquad \ldots \text{(ii)}$$

where B is the price of, say, $2\frac{1}{2}$ per cent Consols; E, the price index of equities; R, the recent change in the rate of change in retail prices; P is the price of bonds; C, the coupon; T, the remaining years to redemption; and K_1 and K_2 are constants. Twenty sets of observations are used in each of the two regression analyses necessary, and these basic data are given in Tables 7.11 and 7.12.

The analysis for equation (i)

It is proposed to go through the analysis for equation (i) step by step; and to give no more than the results for equation (ii). The first step is to convert all the basic data to logarithms and to extract the changes in the rate of change in retail prices on the assumption that investors base their investment decisions on the last six months' change. Since the retail price index records a continuous upward trend throughout the period of 5 years covered by the data, the sign is of no consequence: the greater the rate of change as recorded the greater the fall in the level of bond prices. The correlation is therefore negative, as it is between bond prices and equity prices.

TABLE 7.11 Basic data for equation (i)—monthly averages

	Equities price index (F.T. 500 shares)	*Retail price index (1950 = 100)*	*2½% Consols. Price per £100 (£)*
1965 July	—	178·0	—
1966 January	112	180·5	38·8
April	113	183·3	37·3
July	113	184·2	35·5
October	98	185·4	36·2
1967 January	102	187·1	37·8
April	109	188·8	39·4
July	114	188·3	36·4
October	127	189·0	36·4
1968 January	132	192·1	35·2
April	153	197·1	34·8
July	172	198·3	33·1
October	175	200·0	33·6
1969 January	189	204·2	30·4
April	173	208·3	28·5
July	148	209·0	27·6
October	146	210·8	28·5
1970 January	160	214·4	28·4
April	148	220·1	28·4
July	132	222·9	27·0
October	148	226·2	26·9

Sources: Equities price index, F.T.–Actuaries in *Financial Statistics* (HMSO); Retail price index, Ministry of Employment, *Three Banks Monthly Summary*; Price of Consols, *Financial Statistics.*

The data of Table 7.11 were first converted to four-figure logarithms. Then in order to save arithmetic, the lowest figure in each series was deducted from each figure in the series in the case of the price series of Equities and Consols, and the resulting differences rounded to three decimal places. With retail prices, each four- or five-figure logarithm was reduced by that for the month six months earlier, and then rounded off to three decimal places. It was scarcely worth while to reduce each figure of the resulting series by its lowest figure; so that the unreduced figures were used in the analysis. It should be mentioned that reducing a series by its lowest figure does not affect the absolute variation. Regression analysis by any method depends entirely upon the absolute variation and not the relative variation of the data used in the analysis. Where

TABLE 7.12 Basic data for equation (ii)—
20 January 1971

Stock	*Coupon* (%)	*Remaining life to redemption* (*years*)	*Bond price per £100* (£)
Conversion	6·00	1	99
Gas	4·00	1	97
Exchequer	6·25	1	99
Exchequer	6·75	2	99
Electrical	3·00	2	93
Transport	3·00	2	93
Conversion	5·25	3	94
Treasury	6·75	3	99
Treasury	6·50	5	94
Treasury	6·50	5	95
Exchequer	5·00	7	85
Electrical	4·25	8	78
Treasury	8·50	11	96
Treasury	8·50	15	93
Funding	6·50	16	78
Treasury	5·00	18	64
Funding	5·75	20	69
Funding	6·00	22	70
Treasury	9·00	23	96
Treasury	9·00	23	95

Source: Financial Times, Share Information Service 21 January 1971.

desk calculators are used, reducing the arithmetic by the methods available can save much labour. Where an electronic computer is used, reduction by the lowest figure and rounding off would generally be quite unnecessary. But, as pointed out above, the preparation of data for processing by electronic computer is as important as it is for processing manually. A computer is not a clarifier of basic data.

The resulting series are given in the first three columns of Table 7.13, in the same order as the original data appear in Table 7.11. The key to the symbols used is as follows:

X_2 the log series for E, the price of equities.
X_3 the log series for R, the retail price index (6-monthly changes).
X_1 the log series for B, the price of $2\frac{1}{2}$ per cent Consols.

The remainder of the table contains the squares and products required

	Log E X_2	Log R X_3	Log B X_1	X_2^2	X_3^2	X_1^2	X_2X_3	X_2X_1	X_3X_1
	0·058	0·006	0·159	0·003,364	0·000,036	0·025,281	0·000,348	0·009,222	0·000,954
	0·062	0·011	0·142	0·003,844	0·000,121	0·020,164	0·000,682	0·008,804	0·001,562
	0·062	0·009	0·120	0·003,844	0·000,081	0·014,400	0·000,558	0·007,440	0·001,080
	–	0·005	0·129	–	0·000,025	0·016,641	–	–	0·000,645
	0·017	0·007	0·148	0·000,289	0·000,049	0·021,904	0·000,119	0·002,516	0·001,036
	0·046	0·008	0·166	0·002,116	0·000,064	0·027,556	0·000,368	0·007,636	0·001,328
	0·066	0·003	0·131	0·004,356	0·000,009	0·017,161	0·000,198	0·008,646	0·000,393
	0·113	0·001	0·131	0·012,769	0·000,001	0·017,161	0·000,113	0·014,803	0·000,131
	0·129	0·009	0·117	0·016,641	0·000,081	0·013,689	0·001,161	0·015,093	0·001,053
	0·194	0·018	0·112	0·037,636	0·000,324	0·012,544	0·003,492	0·021,728	0·002,016
	0·244	0·014	0·090	0·059,536	0·000,196	0·008,100	0·003,416	0·021,960	0·001,260
	0·252	0·006	0·096	0·063,504	0·000,036	0·009,216	0·001,512	0·024,192	0·000,576
	0·285	0·013	0·053	0·081,225	0·000,169	0·002,809	0·003,705	0·015,105	0·000,689
	0·247	0·018	0·025	0·061,009	0·000,324	0·000,625	0·004,446	0·006,175	0·000,450
	0·179	0·010	0·011	0·032,041	0·000,100	0·000,121	0·001,790	0·001,969	0·000,110
	0·173	0·005	0·025	0·029,929	0·000,025	0·000,625	0·000,865	0·004,325	0·000,125
	0·213	0·011	0·023	0·045,369	0·000,121	0·000,529	0·002,343	0·004,899	0·000,253
	0·179	0·019	0·023	0·032,041	0·000,361	0·000,529	0·003,401	0·004,117	0·000,437
	0·129	0·017	0·001	0·016,641	0·000,289	0·000,001	0·002,193	0·000,129	0·000,017
	0·179	0·012	–	0·032,041	0·000,144	–	0·002,148	–	–
TOTALS	2·827	0·202	1·702	0·538,195	0·002,556	0·209,056	0·032,858	0·178,759	0·014,115
Correction sums				0·399,596	0·002,040	0·144,840	0·028,553	0·240,578	0·017,190
Sums of squares or products of deviations				0·138,599	0·000,516	0·064,216	0·004,305	–0·061,819	–0·003,075
Averages	0·1413	0·0101	0·0851	$= \Sigma(x_2^2)$	$= \Sigma(x_3^2)$	$= \Sigma(x_1^2)$	$= \Sigma(x_2x_3)$	$= \Sigma(x_2x_1)$	$= \Sigma(x_3x_1)$
Add back	1·9912	–	1·4300						
	2·1325	0·0101	1·5151						

for the least squares equations and for calculating the correlation coefficients. In order to determine the value of K_1, the unattached constant, the averages of the series in the first and third columns need correcting by adding back the lowest figures. The corrections are made at the foot of the columns. To convert the sums of squares and products of the figures in the first three columns to the sums of squares and products of the deviations from the several means, a deduction is made on the lines described in Chapter 6 in connection with the example of Table 6.1.

Only two least squares equations are required to solve for a and b of equation (i). They are:

$$0{\cdot}1386a+0{\cdot}004305b=-0{\cdot}061820$$

$$0{\cdot}004305a+0{\cdot}000516b=-0{\cdot}025655$$

Solved simultaneously, they give:

$$a=-0{\cdot}3522$$

$$b=-3{\cdot}0210$$

Using these coefficients in association with the restored averages to represent the three variables, we have:

$$K_1=1{\cdot}5151+2{\cdot}1325\times 0{\cdot}3522+0{\cdot}0101\times 3{\cdot}0210$$
$$=2{\cdot}2967$$

The complete forecasting equation is therefore:

$$X_1=2{\cdot}2967-0{\cdot}3522X_2-3{\cdot}0210X_3$$

where $X_1=\log B$; $X_2=\log E$; and $X_3=\log R$.

The correlation coefficients

Subscripts 1, 2, 3 . . . are used for indicating particular relationships: thus $r_{1\cdot 2}$ represents the coefficient of correlation between X_1 and X_2. The following are the three simple coefficients in respect of the analysis above:

$$r_{1\cdot 2}=-0{\cdot}656$$

$$r_{1\cdot 3}=-0{\cdot}535$$

$$r_{2\cdot 3}=+0{\cdot}509$$

Of these, the first two are not important, especially as the analysis involves three variables. The third is the more important as a high degree of correlation between independent variables is undesirable: it can give spurious results. From this point of view, the figure of 0·509 can be regarded as moderate: there is little risk of spurious correlation.

Simple correlation is dealt with and demonstrated in Chapter 6. Multiple correlation, i.e. the correlation between the dependent variable and all the independent variables taken collectively, calls for the estimated values of the dependent variable based on the regression equation; and so the subject had to be deferred to this chapter. The multiple correlation coefficient is usually represented by the symbol R, sometimes followed by subscript figures representing the variables taken into account. Thus $R_{1\cdot 23}$ is the coefficient of multiple correlation for X_1 as a function of X_2 and X_3, and $R_{2.13}$ is that for X_2 as a function of X_1 and X_3. It is equal to the standard deviation of the estimated values of the dependent variable divided by the standard deviation of the actual values of the dependent variable. As with the simple coefficients, the multiple coefficient is based on the logarithms of the actual and estimated values where the model is of the logarithmic type.

The estimated value of log B for the set of observations for January 1966 shown in Table 7.11 is 1·5566 arrived at by applying the regression equation to the logarithms of the figures of the equity price and retail price index numbers, that is:

$$2{\cdot}2967 - 2{\cdot}0492 \times 0{\cdot}3522 - 0{\cdot}0061 \times 3{\cdot}0210 = 1{\cdot}5566$$

It will be recalled that the figure of 0·0061 is in effect the log of the ratio of the retail price index for January 1966 to that for July 1965, which is equal to the log of the former less the log of the latter. Similar estimates are made for the rest of the sets of observations, and the standard deviation calculated.

In mathematical terms, the formula for the coefficient of multiple correlation for a problem in three variables is:

$$R_{1\cdot 23} = \frac{\sigma_{1(23)}}{\sigma_1}$$

Substituting for the two standard deviations in respect of the above analysis, we have:

$$R_{1\cdot 23} + \frac{0{\cdot}0395}{0{\cdot}0566}$$

$$= 0{\cdot}698$$

which, as one would expect, is rather higher than either of the two individual simple coefficients. A sign cannot be ascribed to the coefficient of multiple correlation. In any event, it could scarcely serve any useful purpose.

The standard error of estimate

Since there is not much doubt that there is little if any interaction between bond prices on the one hand and either equity prices or retail prices on the other, the relationship expressing the price of Consols as a function of equity prices and retail prices can be said to conform to the theory of error, and so it is possible to calculate the error of estimate without much risk of confounding the assumptions of the theory.

As the estimated values of log B have already been calculated for the purpose of deriving the coefficient of multiple correlation, the first step is to calculate the differences between the estimated value of log B and the actual value for each set of observations. It does not much matter whether the former is deducted from the latter or *vice versa*, the point is that the process should be the same throughout. There will result a series of figures, some with plus signs and the rest with minus signs. In theory, the total of the plusses should equal the total of the minuses; in practice there is often a slight difference due to rounding. Their equality or approximate equality provides a check on some of the arithmetic. It follows that since the total of the series of differences is zero or near zero, their mean is also zero, and the series is therefore its own series of deviations from its mean. The figures of the series, often referred to as Z, are then squared and summed so that the standard deviation can be calculated.

The standard error of estimate is in fact the standard deviation of Z corrected for size of sample, i.e. the number of sets of observations. The standard deviation of Z is:

$$\sqrt{\frac{\Sigma(Z^2)}{n}}$$

And the standard error of estimate is:

$$\sqrt{\frac{\Sigma(Z^2)}{n-m}}$$

where m is the number of variables involved, so that $n-m$ represents the number of degrees of freedom. Where n is large, the effect on the calculation of reducing it by m may be negligible; then the standard error and the standard deviation are nearly identical.

In the example above, $\sum(Z^2)=0{\cdot}02882$, $n=20$ and $m=3$. The standard error of estimate, S, is therefore:

$$S_{1\cdot23}=\frac{0{\cdot}02882}{17}=0{\cdot}0412$$

As was explained in Chapter 3, the standard deviation is a measure expressed in the same terms as the series whose deviation or dispersion it is the measure of. If the series is in tons, then the standard deviation is in tons, and if it is in terms of the logarithms of the sterling value of Consols, then so is the standard deviation, and also therefore the standard error of estimate. The probabilities for the standard error are the same as for the standard deviation. For one standard error, the probability that an observation or actual value will fall within its range is 68 per cent, for twice the standard error, it is 95 per cent.

Usually, the error is expressed as plus or minus after the estimated value of the dependent variable. Thus we have:

$$\text{Estimate of } \log B \pm 0{\cdot}0412$$

For January 1966, we have seen that the estimate of log B is 1·5566 compared with the actual value of 1·5888. The latter falls within the range $1{\cdot}5566 \pm 0{\cdot}0412$. A glance down the Z series shows that 14, or 70 per cent of the 20 actual observations fall within the limits set by one standard error of the estimated values, which provides empirical support for the mathematically derived standard error.

Since a value of log B is a logarithm so also is the standard error. For January 1966, we have:

$$1{\cdot}5566 \pm 0{\cdot}0412$$

Converted to natural numbers, they are:

$$36{\cdot}02 \times \text{ or} \div 1{\cdot}100$$

which for all practical purposes can be expressed as:

$$36{\cdot}02 \pm 10 \text{ per cent}$$

However, since any practical exercise with a logarithmic model is carried out entirely in logarithms until the final conversion, this does no more than demonstrate the significance of a logarithmic standard error.

Practical men sometimes ask, 'What do we do with the standard error when we have it?' It is not an easy question to answer. One may well wonder whether much of the work that has gone into the development and application of error theory is not of academic interest rather than of practical value. However, managers are becoming more and more interested in the degree of confidence they can place in statistical estimates; so that at least some of the work of the academics in this field is not without practical value.

Intrepretation

There are several ways of tackling the problem of interpretation. One, perhaps the most revealing, is to consider the effect of eliminating the influence on the dependent variable of the independent variables, first together, and then each in turn. For this purpose the averages of the independents can be used. We have the complete equation:

$$\log B = 2{\cdot}2967 - 0{\cdot}3522 \log E - 3{\cdot}0210 \log R$$

If we eliminate the influence of log E and log R, all we have left is the unattached constant $K_1 = 2{\cdot}2967$, so that:

$$\log B = 2{\cdot}2967$$

$$\text{and } B = 198{\cdot}0$$

which means that with $\log E = \log R = 0$, the price of $2\frac{1}{2}$ per cent Consols per £100 would be £198 ± 10 per cent. One thing is certain, $2\frac{1}{2}$ per cent Consols never topped £100 for more than a very short time; but then the index of equity prices has never been down to one to give $\log E = 0$. More rewarding, perhaps, is determining the effect of changes in the rate of inflation by seeing how its elimination would affect the price of Consols. We would then have:

$$\log B = 2{\cdot}2967 - 0{\cdot}3522 \times 2{\cdot}1325$$

$$= 2{\cdot}2967 - 0{\cdot}7511$$

$$= 1{\cdot}5456$$

$$\therefore B = £35{\cdot}13 \text{ per } £100$$

compared with the average price (the geometric mean) in the period covered of £32·74 (the antilog of 1·5151). The difference of £2·39 is the fall in price due to an average rise in the rate of increase in retail prices of 2·3 per cent (the antilog of 0·0101 being 1·023). If retail prices became stable, there would necessarily have to be a fall or a series of falls in the rate of increase in prices from the present position. It looks from the evidence above as though the price of 2½ per cent Consols would rise by just over £1 for every fall of 1 per cent in the rate of increase in the retail price level, just as it appears to have fallen by £1 for every rise of 1 per cent in the rate of increase in retail prices.

Similarly for the influence of equity prices: the index is based on 1962=100 when 2½ per cent Consols stood at an average of £41·8. Ignoring the effect of retail prices and using 100 for E, i.e. 2 for log E, we have:

$$\log B = 2{\cdot}2967 - 0{\cdot}3522 \times 2$$

$$= 1{\cdot}5923$$

$$\therefore\ B = £39{\cdot}11 \text{ per } £100$$

which is well within 10 per cent of the actual figure of £41·8. We have then:

when $E = 100, \quad B = 39{\cdot}1$

and when $E = 135{\cdot}7, \quad B = 32{\cdot}7$

Other things being equal, it therefore looks as though the price of 2½ per cent Consols falls by £1 for every rise of five points in the F.T.–Actuaries index of equity prices. This, of course, gives merely a very rough idea of the relationship. Strictly, to conform to the exponential model, logarithms should be used for expressing such a linear relationship.

Between January 1969, when equity prices reached their peak, and October 1970, the index fell from 189 to 148; but 2½ per cent Consols also fell from £30·4 to £26·9. The reason for this seems to be that other things were far from equal: the rate of increase of retail prices was rising rapidly in 1970, and the post-incomes policy wages explosion was threatening still greater rises. Between January and October 1969, the retail index rose by 6·6 points, and in the same period of 1970, it rose by 11·8 points.

The results for equation (ii)

It will be recalled that equation (ii) reads:

$$\log P = e \log C + d \log T + K_2$$

where P is the price of the Government bond; C, the coupon; and T, the remaining life to redemption, and c, d and K_2 are constants. With X_1 for log P, X_2 for log C and X_3 for log T, the following are the averages and sums of squares and of products:

Averages (restored):

$$X_2=0{\cdot}7654 \therefore \text{GM of } C=5{\cdot}83$$
$$X_3=0{\cdot}7523 \therefore \text{GM of } T=5{\cdot}65$$
$$X_1=1{\cdot}9479 \therefore \text{GM of } P=88{\cdot}69$$

Sums of squares and products:

$$(x_2^2)=0{\cdot}366{,}742$$
$$(x_3^2)=4{\cdot}609{,}342$$
$$(x_1^2)=0{\cdot}064{,}637$$
$$(x_2x_3)=0{\cdot}627{,}727$$
$$(x_2x_1)=0{\cdot}027{,}143$$
$$(x_3x_1)=-0{\cdot}329{,}411$$

The following are the results of the analysis:

$$c=0{\cdot}2560$$
$$d=-0{\cdot}1063$$
$$\log K_2=1{\cdot}8319$$
$$r_{1.2}=0{\cdot}176$$
$$r_{1.3}=-0{\cdot}546$$
$$r_{2.3}=0{\cdot}483$$
$$R_{1.23}=0{\cdot}809$$
$$S_{1.23}=0{\cdot}0364, \text{ i.e. } \pm 8{\cdot}7 \text{ per cent}$$

Of the 20 Z values, only 12 or 60 per cent in this case fall within the standard error of 0·0364. An explanation of the remarkably low value of $r_{1\cdot 2}$ has been offered above. The standard error is disappointingly high too. Perhaps it would have been better to limit the analysis to the more active stocks, which would have meant excluding all those with a coupon of less than 4 per cent; but this would have had the undesirable effect of reducing the range of variation of the coupon series.

Combining the two equations

In the natural forms of equations (i) and (ii), the two unattached constants K_1 and K_2 appear as multipliers:

$$B=E^aR^bK_1 \qquad \ldots \text{(iA)}$$
$$P=C^cT^dK_2 \qquad \ldots \text{(iiA)}$$

With equation (iiA), the need for a multiplier is obvious: it provides a measure of the state of the market. The price trends of any undated stock provides a measure of the movement of the market, since Government stocks tend to change *pari passu* in price, apart from the effect of shortening life at the shorter dated end of the market. As calculated, K_2 appears as a constant reflecting the state of the market at some particular time (20 January 1971 in the example). To reflect changes in the state of the market, K_2 needs to be regarded as partly a variable. As a variable it is a function of the price of any undated stock. Let the undated stock be $2\frac{1}{2}$ per cent Consols: then we have:

$$K_2 = f(B)$$

Since B is a function of E and R, it can be said that $K_2 = f(E, R)$. However, it would clearly not be logical to substitute the completed form of equation (i) for log K_2 in equation (ii), since K_2 is not directly related to Consols or any other particular stock. What seems to be necessary is an adjustment of the value of log K_1 so that equation (i) is equal to log K_2. The adjustment is equal to the difference between log K_2 and log B. The former is 1·8319 and on 20 January 1971 the latter was 1·4253, an excess of the former over the latter of 0·4066, the required addition to log K_1, which becomes 2·7033 for the purpose.

The complete forecasting equation in its logarithmic form is:

$$\log P = 2{\cdot}7033 - 0{\cdot}3522 \log E - 3{\cdot}0210 \log R + 0{\cdot}2560 \log C - 0{\cdot}1063 \log T$$

ELECTRONIC COMPUTING

A word about using a computer for the work of regression analysis seems to be worthwhile. The whole of the foregoing analysis has been carried out manually in order that the several stages in the work can be demonstrated, and to show how the problems and difficulties that arise can be ironed out.

It seems that many manufacturers of electronic computers have programmes ready made for least squares regression analyses in any number of variables. A ready made programme for the appropriate number of variables would save programming time. A print-out of the various values, as given above for equation (ii), is useful for enabling the analyst to exercise an internal consistency check on the basic data, the accuracy of the punching work, and the calculations.

One of the advantages of using a computer is that it can handle much larger numbers of sets of observations than can be readily handled by a desk calculator. It can also handle a large number of variables more easily. For an analysis of the gilt-edged market, it could take into account all four variables in a single analysis without difficulty, and use hundreds of sets of observations for the purpose. This would give the analyst much more satisfactory results, and more time for him and his staff to devote to data preparation, model building and testing the regression behaviour of his data by applying to them one or more of the less sophisticated methods of regression analysis. Some of his preliminary work could be done stage by stage on the computer before a least squares analysis is attempted.

Another advantage of using a computer is that less rounding off is necessary. Seven-figure logarithms, for instance, could be used for analyses based on an exponential model; though since the squares and products would run to 14 decimal places, some rounding off might be necessary or at least convenient at this stage.

Whether the methods of saving arithmetic demonstrated above could be used with advantage on a computer is a matter for the programmer to decide in consultation with the statistical analyst and specifications writer.

CONCLUSION

Experience shows that of all the methods of statistical analysis, regression analysis is without doubt the most valuable as an *open sesame* to management information—those treasures that lie buried in the kind of statistical data which all large and medium sized companies and some smaller ones now collate as a matter of routine for planning and control purposes. For my part, I have laid bare many invaluable secrets thus hidden away, by applying the methods of regression analysis in the fields of market demand, costing, manpower and production.

Regression analysis is a big subject. The leading authority on it is Mordecai Ezekiel, whose work *Methods of Correlation Analysis* (Wiley, First edition, 1930), should be read and studied by all industrial statisticians and management accountants. Many books, monographs and articles have been written and published dealing with some particular real-life analysis by a regression method. Some of these are well worth reading, dealing as many of them do with some aspect of regression analysis in detail—the model, the method used, the errors in the calculated parameters and such like. Others, however, are not so good;

they fall into one trap or another: lack of preparation of the statistical data used, the blind adoption of the linear assumption or even of a quadratic form of model without discussing or giving any reason first, a misinterpretation of the parameters through lack of a hypothesis. A rational, scientific, systematic approach is as essential in applying an advanced method of statistical analysis such as the methods of regression analysis as it is in applying a medical, a surgical, an agricultural or a management technique.

8

Deriving demand elasticities

Demand elasticities are of fundamental importance in sales forecasting, advertising policy, price fixing and in company financial policy generally. The best known and most useful is the price-elasticity of demand, sometimes called the *elasticity of demand with respect to price.* Other demand elasticities of practical value are those with respect to national income, to the price level of competitors brands, to the company's own advertising and sales promotion expenditure, and to competitors' advertising expenditure; there is almost no end to the list of factors or potential factors in the demand for the company's brand of the product or service.

Price and its own advertising and sales promotion are the only factors that a company has control of. Both are subject to the law of diminishing return, advertising more obviously than price; for with price, it depends upon what is meant by *return.* If it is quantity sold, then as price falls, the quantity increases but by diminishing amounts for every unit decrease in price. If it is sales proceeds, then from a relatively high price, sales proceeds rise also at a diminishing rate, until they fail to record an increase at all, and ultimately turn downward as price continues to fall. If it is profit, then as will be gathered from the conclusions of Chapter 11, the law applies in both directions: either a rise or a fall in price from its optimum, will result in a diminishing profit. In this chapter, we are concerned with gross revenue (i.e., sales proceeds) and quantity sold, and not at all with profit which calls for the introduction of annual cost into the equation.

Something has already been said in Chapter 5 about the elasticities of demand with respect to price and other factors. The relationship between theory and practice in model building was demonstrated. Here, it is proposed to go more deeply into the relationship between price and quantity sold. The economic law of demand states that a rise in price causes a fall in demand and a fall in price causes a rise in demand; that is, in statistical terminology, the two are negatively correlated. The law gives rise to a number of questions: What is the shape of the law? What is meant by price? Does the law not assume that purchasers and potential purchasers behave rationally, that they are knowledgeable about quality? It is proposed to examine these questions in the following sections.

THE PRICE-DEMAND MODEL

Although demand is more often a function of price than price is of demand, economists invariably represent the law graphically as though price were a function of demand, with price (or gross revenue) measured on the vertical axis and demand in terms of quantity measured on the horizontal axis. As it happens, the convention has the important advantage of convenience. Nevertheless, the basic equation for analytical purposes expresses quantity sold as a function of price (and the various other factors that are taken into account).

Theoretically, the shape of the demand law or model is something of a hybrid. It is neither linear nor logarithmic, but somewhere in between, with both the theoretical evidence and the empirical evidence supporting the view that it lies much closer to the logarithmic type than to the linear. If it were linear, we should have

$$p=b-aQ \qquad \ldots \text{(i)}$$

where p is the price; and Q, the quantity sold in unit time. Figure 8.1 shows the graph of such an equation. It is a straight line falling to the right from a point equivalent to b on the vertical axis to a point on the horizontal axis where $p=0$, that is, where $aQ=b$ and $Q=b/a$. That a

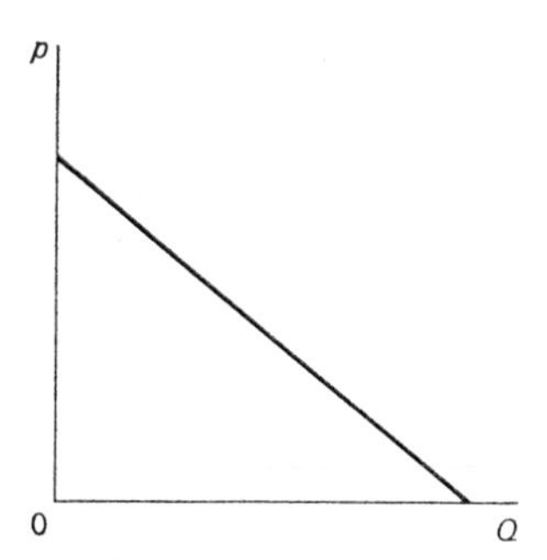

FIG. 8.1 Linear price graph

price equal to b should reduce demand to zero, and that zero price should induce a demand equal to b/a in unit time are ideas that seem to be theoretically indefensible.

However, let us consider the alternative and more practical approach to the theory of demand, in which sales proceeds, S, in unit time are expressed as a function of demand Q. From equation (i), we have:

$$S = pQ = bQ - aQ^2 \qquad \ldots \text{(ii)}$$

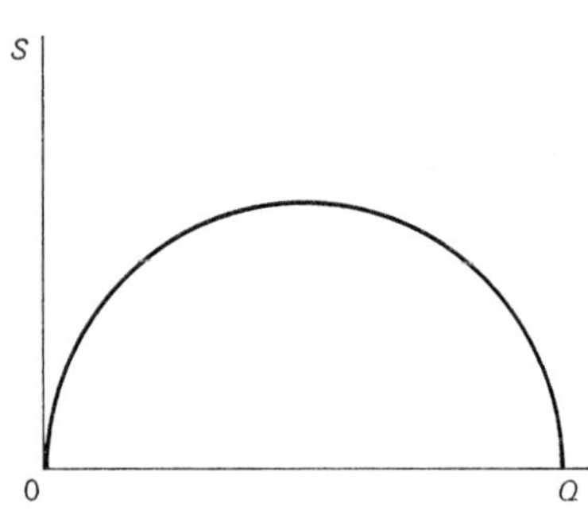

FIG. 8.2 Theoretical sales graph

whose graph is a parabola passing through the origin and concave downward as in Fig. 8.2. The peak is reached where the differential coefficient of S with respect to Q is zero:

$$\frac{dS}{dQ} = b - 2aQ$$

Equating to zero, we have for peak of graph:

$$2aQ = b$$

$$\therefore Q = \frac{b}{2a} \qquad \ldots \text{(iii)}$$

Substituting for Q in equation (i), we have:

$$p = b - \frac{ab}{2a}$$

$$= b - b/2$$

$$= b/2 \qquad \ldots \text{(iv)}$$

which is the price that would maximise the gross revenue, S. Since

$S=pQ$, its maximum value would be:

$$S=pQ=\frac{b}{2a}\cdot\frac{b}{2}$$

$$=\frac{b^2}{4a} \qquad \ldots \text{(v)}$$

Oddly enough, the general theoretical shape of the true gross sales graph is parabolic, i.e., it rises from the origin to the right, is concave downward and reaches a peak, whence it falls to the right, finally reaching the horizontal axis. Strictly, then, it seems the quadratic form might be appropriate to the gross revenue function; but since it assumes that the price-demand function is linear, it must be regarded as invalid. In the early days of attempts to derive the price-elasticity of demand for particular products, analysts often used a linear price-demand function as their model, but that was before economists and statisticians had come to realise the importance of having a valid model to serve as a stepping stone between hypothesis and statistical analysis, or even to realise that there could be anything but a linear form of equation as a model.

The logarithmic form

We now turn to consider the logarithmic form of the price-demand and gross revenue model. Here, both graphs are curvilinear: the former falls from an uncertain height for very high prices and low demand, is concave upward, and tends ultimately to become asymptotic to the horizontal axis for very low prices and high demand (see Fig. 8.3); and

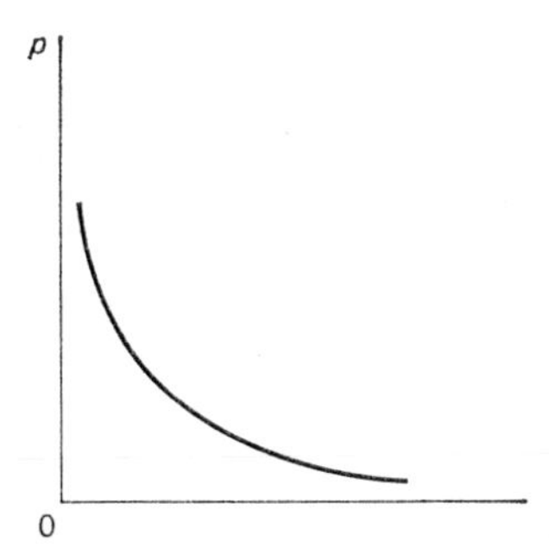

FIG. 8.3 Constant elasticity price graph

the latter rises from the origin for very high prices and low demand, is concave downward and like the price-demand graph, also tends to become asymptotic to the horizontal axis for very low prices and high demand (see Fig. 8.4).

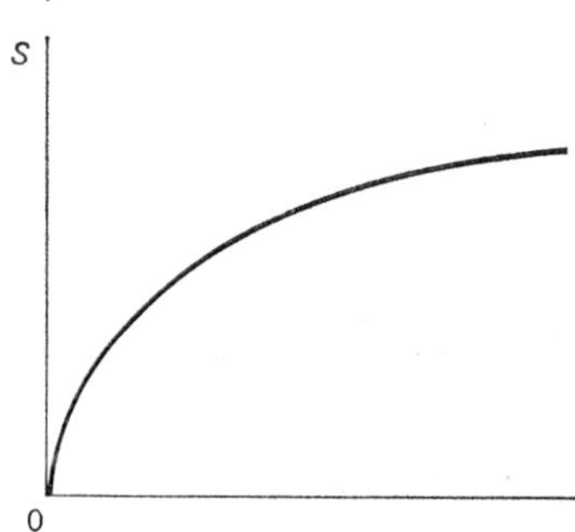

FIG. 8.4 Constant elasticity sales graph

The corresponding equations are:
For the price-demand graph of Fig. 8.3:

$$Q = kp^{-e} \qquad \ldots \text{(vi)}$$

i.e.

$$p = KQ^{-1/e} \qquad \ldots \text{(vii)}$$

and for the gross revenue graph of Fig. 8.4.:

$$S = pQ = KQ^{1-1/e} \qquad \ldots \text{(viii)}$$

where k and K are constants, K being equal to $k^{1/e}$, and e is also a constant, representing the price-elasticity of demand. The advantage of having the price-elasticity of demand represented by a constant can scarcely be exaggerated; and although this must have some weight in deciding the appropriate type of model, it should not be allowed to be decisive.

It should be kept in mind that the four graphs in Figs. 8.1 to 8.4 do not in any sense represent time trends. They represent *rates* of sales by quantity and, in Figs. 8.2 and 8.4, by money value as well as in unit time. It is best to think of them as giving the *annual* rates of sales at a *point of time.*

The empirical evidence

There is a modicum of empirical evidence to support the theory that the elasticity of demand is a function of price—that it tends to fall as price falls. However, the evidence is that the elasticity of demand is slow to change with movements of price, so much so, indeed, that the effect on it of a doubling or halving of price is so insignificant that it can be regarded as a constant for all practical purposes. It is true that the published evidence applies to whole markets, such as those for motor cars, refrigerators, sugar, flour and tea, rather than to the markets for brands or makes of them; but there is no reason for supposing that it does not apply equally to brands and makes. If there is a reason, it could be sought in the low level of the elasticity of demand for a commodity as a whole relatively to that for any brand or make of it. The elasticity of demand is a measure of the competition encountered by a commodity or brand of it. The existence of close substitutes more than any other factor creates competition. Brands have close substitutes, a commodity as a whole has not. Therefore the market in brands is much more competitive and the elasticity of demand higher than the market in commodities as a whole.

It is now appropriate to return to the graph of Fig. 8.2 in order to reconsider it. It can be read, not as the graph of a quadratic equation, but as representing the general shape of the complete gross revenue curve, which lies somewhere between a quadratic and a logarithmic with characteristics of the former rather than of the latter. Where the graph reaches its peak, a change in price does not affect the rate of gross revenue, any such change being entirely offset by a change in quantity sold in the opposite direction. In practice, the peak takes the form of a plateau where the price-elasticity of demand is unity. Equation (viii) reads:

$$S = KQ^{1-1/e}$$

At the peak, the differential coefficient of S with respect to Q is zero:

$$\frac{dS}{dQ} = (1 - 1/e)KQ^{-1/e} \qquad \text{. . . (ix)}$$

which is zero when $e = 1$. It also follows that the price-elasticity of demand exceeds unity for the arc of the graph of Fig. 8.2 lying to the left of the peak, and falls short of unity for the arc to the right of the peak.

Many commodities as a whole, probably most of them, have price-elasticities of demand that are less than unity, whereas it is true to say that all brands and makes of commodities have individual price-elasticities of demand that are greater than unity. For whole commodities, then, the gross revenue lies below its maximum because prices are on the low side, whereas for brands, it lies below its maximum because prices are on the high side. Competition amongst brands and makes accounts for both the former and the latter, or more precisely, for the relatively low prices of whole commodities, and for the relatively high price-elasticities of demand for individual brands. Another factor in the latter is the marginal cost of production, whose influence on brand pricing forms a subject of Chapter 11.

An alternative to the graph of presenting the theory of demand is the sales schedule showing the price-elasticity of demand for each price. In mathematical terms, the price-elasticity of demand, e, is:

$$e=-\frac{dQ}{dp}\cdot\frac{p}{Q} \qquad \ldots \text{(x)}$$

Using equation (vi), we have:

$$\frac{dQ}{dp}=-ekp^{(-e-1)}$$

so that

$$\frac{dQ}{dp}\cdot\frac{p}{Q}=-ekp^{(-e-1)}\cdot\frac{p}{Q}$$

Substituting for Q its value in equation (vi), we have:

$$\frac{dQ}{dp}\cdot\frac{p}{Q}=-\frac{ekp^{-e}}{kp^{-e}}$$

$$=-e$$

The formula of equation (x) is of general application. For the straight line price-demand equation, we have from equation (i):

$$aQ=p-b$$

$$\therefore\ Q=p/a-b/a$$

Since a and b are constants, we can rewrite this:

$$Q=cp-f \qquad \ldots \text{(xi)}$$

$$\therefore\ \frac{dQ}{dp}=c$$

and

$$\frac{dQ}{dp}\cdot\frac{p}{Q}=\frac{cp}{Q}$$

so that for a linear price-demand equation:

$$e = -\frac{cp}{Q} \qquad \ldots \text{(xii)}$$

A theoretical demand schedule

Table 8.1 contains a complete purely theoretical demand schedule and the price-elasticities of demand, for the linear price-demand equation:

$$Q = 100 - 10p$$

It begins at the origin of its graph with zero quantity and a price-elasticity of demand of infinity and finishes on the horizontal axis with zero price for zero price-elasticity of demand.

TABLE 8.1 A complete theoretical demand schedule

Price p £	*Quantity sold* Q *No.*	*Gross revenue* $pQ=S$ £	*Price-elasticity of demand* $e = -10p/Q$
10	0	0	infinity
9	10	90	9·000
8	20	160	4·000
7	30	210	2·333
6	40	240	1·500
5	50	250	1·000
4	60	240	0·667
3	70	210	0·429
2	80	160	0·250
1	90	90	0·111
0	100	0	0·000

It will be seen from the third column, which gives the value of S, that the gross revenue reaches its peak at £250 for a price of £5, and a price-elasticity of demand of unity.

It goes without saying that a complete schedule cannot be built in practice. Apart from the fact that the elasticity of demand changes very much more slowly with changes in price than the table suggests, no brand or commodity ever went through the whole range of prices from zero upward. The schedule is a fiction designed to help the reader to a fuller understanding of the price-elasticity of demand and its place in the theory and practice of demand analysis.

OTHER ELASTICITIES OF DEMAND

Price can be regarded as the principal independent variable where the quantity sold in unit time is the dependent variable adopted for analytical purposes. As such, its relationship with quantity must of necessity set the pattern of the complete model. In short, if we decide that the relationship conforms more nearly to the logarithmic than to any other practicable type then the complete model must also be logarithmic. Hybrid models are difficult and often impossible to handle.

However, this is not to say that there would be no profit in examining the relationship between the adopted dependent variable and the other independents. Where the primary relationship is logarithmic for instance, it may be considered advisable to introduce one or more of the other independent variables as an exponent of ϵ or 10 as in equation (viii) of Chapter 5:

$$Y = K10^{rx}$$

i.e., linearly:

$$\log Y = \log K + rx$$

or as an exponent of r which is appropriate to a series increasing or decreasing in geometric progression as referred to in Chapter 7:

$$Y = Kr^{x}$$

$$\log Y = \log K + x \log r$$

where $r = 1 + R/100$, R being the percentage rate of increase. It should be mentioned that these two equations are the same in effect, for since r is a constant so also is $\log r$ in the linear form of the latter. It is the value of r not $\log r$, that is being sought, so that the former equation is preferable if only because it would be less confusing, for the two equations are so far similar in their logarithmic forms that the value of $\log r$ in the latter would turn out to be precisely the same as the value of r in the former.

It is proposed to take a brief look at the factors other than price which one must usually take into account in the regression analysis of a time series of the quantity in demand.

National income

By national income is meant the appropriate series as published in the *National Income* blue book[1]. Table 1 of the blue book gives a number of series valued at prices current in the year, and Table 14 gives the same series more usefully valued at 1963 prices. The several series that are most likely to be of use in the analysis of market demand are:

A Consumers' expenditure;
B Public authorities' current expenditure on goods and services;
C Gross domestic fixed capital formation;
D Value of physical increase in stocks and work in progress.

Series A covers consumers' non-durable goods only. Personal expenditure on cars, new houses and other durable goods and assets is included in series C. Series D would be of interest from this point of view only to producers of raw materials and products for further industrial processing. The income side of the accounts might be considered more appropriate to the purpose of deriving demand functions. In Tables 2 and 3 of the blue book figures of personal and corporate income are given. Needless to say, no management accountant or industrial economist or statistician can afford to neglect the mine of information contained in the blue book. There are many detailed series providing a breakdown of the figures given in the early tables. Indeed, a careful study of the whole book would be found to be well worth while.

It should be kept in mind, however, that with consumer goods in general, both durable and non-durable, detailed figures may not always be appropriate. It is general welfare, growing or declining affluence, that counts. As real income grows, the total spent on some particular necessity may also grow; but the proportion of the total so spent tends to decline. Today's comforts are tomorrow's necessities; and to-day's luxuries are tomorrow's comforts. Perhaps this may seem to be taking too long a view. A comparison of the figures in Table 23 of the blue book shows this to be quite untrue. Table 8.2 is based on the figures in the 1970 issue of the blue book.

Expenditure at 1963 prices shows the greatest increase between 1959 and 1969 in the running costs of motor vehicles, and the smallest increase in tobacco, with food a fairly close second. The blue book table contains much more detail than that shown for the selected items as

[1] *National Income and Expenditure*, HMSO.

well as many more items. Food alone is broken down into ten items. For the purpose of some statistical analyses, consumers' expenditure at current prices would be more suitable. Table 22 of the blue book gives such figures in similar detail to that of Table 23.

TABLE 8.2 Consumers' expenditure at 1963 prices: selected items

	1959	*1969*	*Increase*
	£m.	*£m.*	%
Total	17,747	22,629	27·5
Food	4,423	4,921	11·3
Alcoholic drink	1,023	1,368	33·7
Tobacco	1,231	1,250	1·6
Housing	1,924	2,608	35·6
Clothing	1,641	2,123	29·4
Durable goods	1,314	1,696	29·1
Recreational goods	330	505	53·0
Motor-vehicle running costs	438	1,134	158·8
Expenditure abroad	259	303	17·0

Table 8.2 shows the extent to which the pattern of consumer expenditure in real terms can change in the course of ten years. A little reflection on the figures suggests that the relationship between consumers' income and consumers' demand for individual items is logarithmic rather than linear. For the necessaries of life, the income-elasticity of demand is less than unity, the shape of the graph being that of Fig. 8.5. For comforts, the income-elasticity of demand hovers about unity, the shape of the graph being a fairly straight line rising from the origin

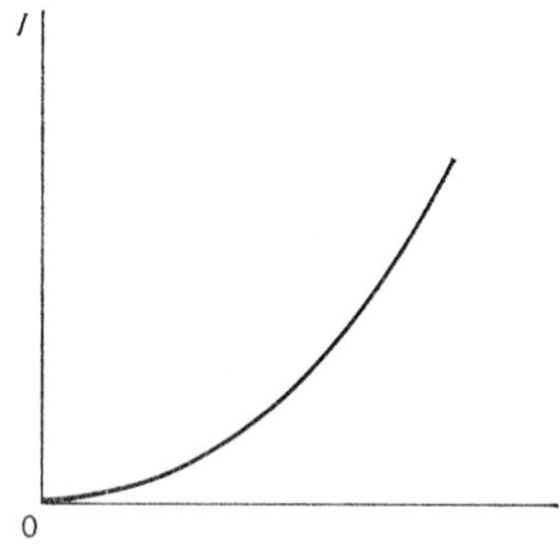

FIG. 8.5 Income giving diminishing returns

to the right. And for luxuries, it is above unity, the shape of the graph being like that of Fig. 8.6. As income *per capita* grows, a luxury becomes a comfort: and a comfort a necessity, so that one could imagine the graph of Fig. 8.6 tending to straighten out, and to curl upward like the graph of Fig. 8.5. Again, it is necessary to point out that graphs of this type do not represent time trends. They present relationships between averages of entities such as price or unit cost and rates in unit time of such entities

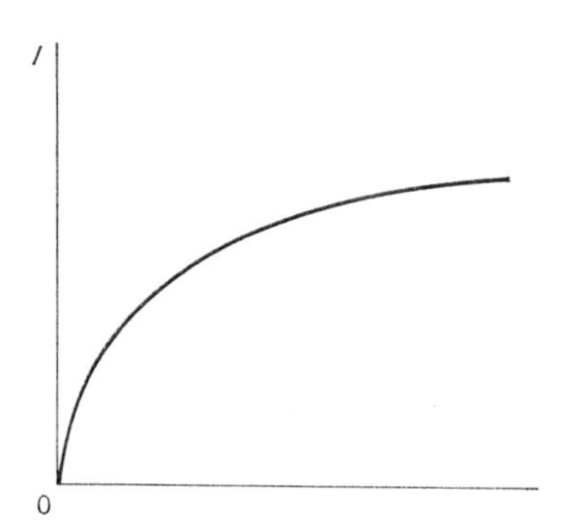

FIG. 8.6 Income giving increasing returns

as sales, income and advertising expenditure. The table also suggests that a regression analysis of market demand based on time series over a period of four years or more would need to take account of time as an independent variable, especially where the analysis applies to commodities or services which show appreciable growth, unless a specific growth factor can be found and embodied in the analysis. In spite of rising prices, the consumer demand for petrol has expanded very appreciably in recent years. Here we have an example of a growth factor in the number of cars licensed, for which statistics are available, so that in an analysis of the market demand for petrol, lubricants, tyres sold separately or vehicle repair services, there would be no need to take account of time as an independent variable.

Advertising

Advertising and sales promotion yield to the law of diminishing returns, as suggested above. With the rate of advertising expenditure measured on the vertical axis and quantity sold on the horizontal axis, the graph rises to the right with an upward concavity, each unit increase in the rate of sales calling for a greater advertising effort. The shape of the graph is the same as that in Fig. 8.5.

Competitors' advertising and sales promotion have the opposite effect on the sales of the company's product, but since expenditure on them likewise yields to the law of diminishing return from the competitors' point of view, the relationship between it and the company's own sales is also likely to be exponential in type. The shape of the graph would be like that of Fig. 8.3—it would fall to the right with an upward concavity.

Competitors' prices

Changes in the prices of competitors' brands have the opposite effect on the company's own sales to changes in the same direction in the company's price of the product. The shape of the graph of the price level of competitors' brands against the sales of the company's product, is likely to be similar to that of the graph depicted in Fig. 8.5.

The purchasing power of money

There are two ways of looking at other goods and services: one is that they are counter-spending attractions like competitors' brands, and the other is that changes in their price level provide an inverse measure of the purchasing power of money. If one sees them in the light of counter-spending attractions, there would be a tendency to adopt their price level as an independent variable. On the other hand, if one sees their price level as an inverse measure of the purchasing power of money, there would be a tendency to deflate all statistics expressed in terms of money of the variables used by reference to their price level.

There is little to be said in favour of the former point of view. Many goods and services compete only with other brands of the same product. This applies more particularly to investment goods and industrial services. The only substitute for a potters' wheel is another potters' wheel, and when a farmer thinks of engaging a contractor to plough his arable, he would not consider the alternative possibility of spending his working capital on consumer goods and services. But there is also a wide range of consumer goods and services, consisting for the most part of the necessaries of life, which scarcely compete with other goods and services outside the product itself. It is amongst luxury goods and services, which have a high price-elasticity of demand, where one finds a wide range of competition. Luxuries compete with one another as counter spending attractions as well as brand against brand. There may thus be a case for regarding the general price level as an independent variable where the demand for a luxury is to be analysed; but it is not a strong case.

Since the be-all and end-all of economic activity is the satisfaction of human wants, the only acceptable measure of the purchasing power of money is the inverse of the price index of consumer goods and services as a whole. In order to express prices and other money values in terms of a stable currency, it is necessary to multiply them by the index of the purchasing power of money, which is the same as dividing them by the retail price index. This would do away with the use of the purchasing power of money as an independent variable, and thus incidentally save one factor in the analysis. It would also show how the price of the brand has changed relatively to changes in the prices of other goods and services. Some producers maintain the money prices of their goods at a stable level for long periods. In times of inflation, the prices of such goods are falling in terms of other goods and services, and their correction by reference to the retail price index gives the series of prices their value in real terms.

The difficulties inherent in the correction of company accounts for inflation where the choice of a standard of comparison must necessarily be arbitrary, are not encountered in the statistical field. The price of the company's brand may be deflated by reference to a retail price index based on, say, 1963=1·00; and personal income, by reference to a retail price index based on another year. The process may seem arbitrary and inconsistent, but it does not affect the results of a regression or similar analysis in the least. The important point to watch is that the estimates to be used when forecasting the demand for the firm's product embody the same albeit inconsistent corrections.

Size of market

To some extent changes in the size of the market for the brand of a commodity or service are taken care of by price, and advertising expenditure. For the rest, where a brand enjoys a limited area of the country as its market, market expansion may take place spontaneously, without any special effort on the part of the company. There is no doubt that such a phenomenon is not taken care of by price or income or any other factor, yet it could result in an appreciable increase in demand.

There seem to be two ways of accounting for it in a regression analysis: one is by counting it as a separate independent variable, with the population of the area of the market as its measure; and the other is to adjust the national income figures to area income figures by reference to the proportion of the total population covered.

It is a difficult problem, mainly because population is scarcely a good measure of the size of a market when the market is tending to grow. Demand in a new area is scarcely likely to be so intense as in an old area. Perhaps it is not so much that population is a poor measure of the size of the market as that sales resistance and brand loyalty may be factors to be reckoned with when a market is expanding terrestrially.

Unquestionably, without these disturbing factors, demand would be a linear function of changing market size, other things being equal. And if market size is taken into account by adjusting the national income series in the way suggested above, it would tend to straighten the relationship between income and demand. However, secondary factors of this kind must not be allowed to dictate the type of demand model to be used.

SUMMARY

Of the several elasticities of demand considered above, those in respect of price and competitors' advertising turn out to have negative signs as a rule. The remainder have positive signs. Figure 8.3 gives the normal shape of graph for functions with negative elasticities, and Figs. 8.5 and 8.6 give the normal shape of graph for functions with positive elasticities.

It should be mentioned that when the price-elasticity of demand has a negative sign, it is said to be positive, presumably because a negative sign is normal. When it has a positive sign, that is, where demand increases when price rises—a phenomenon that we shall be examining later in this chapter—the elasticity is said to be negative. It is all very confusing to the non-mathematician, but such is the convention, which may be perfectly logical to the mathematician. In any event the convention is now so deeply rooted in the literature of management finance, economics and statistics that any attempt to change it would be pointless.

DERIVATION

It will have been gathered that the standard method of deriving demand elasticities is by the application of regression analysis to time series.

Time series are not always available; there are none for new products for instance. However, the most important elasticity is that with respect to price, since as will be shown in Chapter 11, it is one of the two values required for rational pricing. The elasticity of demand with respect to the company's own advertising and sales promotion is also important. Both price and advertising unlike the other factors in demand, are under the direct control of the company.

With neither of these two elasticities of demand is it necessary to adopt a hit or miss policy even for a new product, once it has settled down in the market. Expert opinion together with a little arithmetic can produce estimates of both elasticities, and though they may be far from accurate, they are better than nothing as a guide to pricing and advertising policies. Expert opinion will be found amongst the sales staff, the marketing men including the research workers, and the advertising executives.

In order to demonstrate the procedure, a simple hypothetical example can be used. First, the experts on sales are selected, and they are each invited to answer a simple questionnaire on the following lines: 'In September, the Company sold 122,000 food mixers of our model XYZ to retailers at an average price of £21. How many do you think we should have sold if the price had been £18, and how many if it had been £25?' As many knowledgeable people as possible should be invited to give estimates, each independently of the others. Responders should be asked not to collaborate or to discuss the question with other people. With all the returns in front of him, the statistician must then make up his mind whether to average the answers as they stand or whether to call a meeting of the responders to discuss the answers and to arrive at agreed figures.

If he decides to average the answers as they stand, he must then make a choice between the arithmetic mean and the geometric mean; or he could choose to use both and compare the results. Strictly, it seems that as the relationship between price and demand is exponential, he should prefer the GM to the AM. Suppose the two averages are 149,000 and 97,000 respectively, then the necessary calculations would be as given in Table 8.3.

Before an attempt is made to interpret the results of the analysis, a word about the negative sign seems called for. As stated above, the price-elasticity of demand in normal circumstances where an increase in price causes a fall in demand, as in the example, is expressed as a positive figure, although mathematically, it always has a negative sign.

In the example, for instance, it would be stated as averaging 1·309, not −1·309.

The figure of 1·309 is a weighted average of the other two. Estimates for two different postulated prices, one above and the other below the actual, are obtained to provide an internal consistency check on the sales people's estimates of sales. Since the two estimates differentiated from the actual give much the same solution price-elasticity (1·300 compared with 1·316) it can be said that the check shows the estimates to be internally consistent. It may be taken that the price-elasticity of demand for the company's XYZ model is somewhere in the region of 1·3, a little higher on the evidence, but since sales people the world over tend to overestimate the effect on sales of price-change, possibly rather lower. The wise statistician would, of course, examine his basic data in detail. One thing he should do is to calculate the standard deviation of the estimates of quantity for each of the two postulated prices, and derive separate solution price-elasticities based on the upper and lower limits.

With the company's own advertising and promotions expenditure, the elasticity of demand has a positive sign, and unlike the price-elasticity with a positive sign, it is spoken of as positive elasticity. It would be very bad advertising indeed that had a negative elasticity of demand. To derive it by recruiting expert opinion, the same procedure is adopted. There may be a minor problem of time lag, but the questionnaire could easily be framed to provide for it. 'Last year, our expenditure on advertising and promotions in respect of our XYZ food mixer was running at £10,000 a month. In September, the company sold 122,000 of the model to retailers. How many do you think we should have sold if advertising and promotions expenditure had been running at £8,000

TABLE 8.3 Deriving the price-elasticity of demand from estimates

Price			*Quantity*			*Price-elasticity of demand*
Postulated or actual p	$\log p$	$\Delta \log p$	*Estimate or actual* Q	$\log Q$	$\Delta \log Q$	$\frac{\Delta \log Q}{\Delta \log p}$
18	1·2553	—	149,000	5·1734	—	—
21	1·3222	0·0669	122,000	5·0864	−0·0870	−1·300
25	1·3979	0·0757	97,000	4·9868	−0·0996	−1·316
		0·1426			−0·1866	−1·309

a month, and how many if it had been running at £12,000 a month?' Suppose the replies averaged 102,300 and 140,700 respectively. The same method of analysis—that of finite differences—is applied to these basic data to obtain the solution advertising-elasticities of demand. Table 8.4 contains the calculations.

TABLE 8.4 Deriving the advertising-elasticity of demand from estimates

Advertising expenditure			*Quantity*			*Advertising-elasticity of demand*
Postulated or actual A	$\log A$	$\Delta \log A$	*Estimate or actual* Q	$\log Q$	$\Delta \log Q$	$\frac{\Delta \log Q}{\Delta \log A}$
8,000	3·9031	—	102,350	5·0101	—	—
10,000	4·000	0·0969	122,000	5·0864	0·0763	0·787
12,000	4·0792	0·0792	140,700	5·1482	0·0618	0·780
		0·1761			0·1381	0·784

Since the two individual estimates of the advertising elasticity (0·787 and 0·780) are much the same, the average expert opinion satisfies the internal consistency test, so that the average elasticity of 0·784 may be accepted for the time being, until more actual data become available. It also satisfies another test that can be applied, one based on the theory that advertising expenditure conforms to the law of diminishing returns. The test is the magnitude of the elasticity of demand. That it is less than unity shows that the experts' estimates conform to the law. If it had been above unity, then the law of increasing return would be satisfied. It is possible but highly improbable that advertising could ever conform to increasing returns. Where the estimated advertising-elasticity does indeed turn out to be above unity, the statistician should analyse the returns individually and ask those who have given estimates showing an increasing return to try again.

Just as sales and marketing people tend to overestimate the effect on demand of changes in price, so advertising men (*ad men*, in the jargon) tend to overestimate the effect on demand of changes in advertising expenditure. A good industrial statistician is a student of bias in estimates, and he may thus have a good idea of the degree to which he should discount individual estimates of demand for postulated prices and advertising expenditure.

With the price and advertising elasticities of demand, the statistician has gone a long way towards deriving a complete stochastic demand function for the brand. The elasticity of demand with respect to national income usually follows the price-elasticity: when one is high, the other is high and *vice versa*; but they can rarely be the same. The price-elasticity of demand normally stands above unity. The income-elasticity can be above or below unity according as the product is a luxury or a necessity. Experience with other brands of the same product may indicate the order of magnitude of the figure to adopt for the time being. The same applies to the elasticity of demand with respect to competitors' advertising on their brands of the product. It could well be that competitors enjoy the same elasticity of demand with respect to advertising expenditure as the company does. Their advertising expenditure yields a diminishing return to themselves, and therefore a diminishing loss to the company itself.

More complex is the elasticity of demand with respect to the price level of competitive brands. Undoubtedly, competitors find that much the same price-elasticity of demand applies to *each* of their brands of the product as the company does to its own brand. But the price-elasticity is theoretically somewhat lower for competitors' brands as a whole, since competitors as a whole have only one competitor in the market, the company itself. If could be below unity, as it probably is where the price-elasticity of demand for the whole product is very low, say, 0·5 or less, unless the company itself has a very large share of the market, say, at least 40 per cent.

THE NEGATIVE PRICE-ELASTICITY OF DEMAND

That a reduction in price causes a rise in demand, and an increase in price causes a fall is orthodox theory, and until recently at any rate, the academically respectable theory. Now, since Dr Alfred Oxenfeldt, Professor of Marketing at the Columbia Graduate School of Business and an authority on pricing policy, expressed the view that with some products in some circumstances, one sure way to raise demand is to increase the price, research workers at other schools of business have gone out in to the field to find out for themselves the kinds of goods that appear to enjoy a negative elasticity of demand.

The general conclusion drawn from the field work seems to be that where the quality of goods is difficult to test, purchasers and potential purchasers tend to judge it by reference to price; the higher the price the better the quality. 'You only get what you pay for' is a popular saying. Yet many who quote it would readily admit that price is not a perfect test of quality: that one may pay more for a lb of sausage of the same quality as sausage that one could buy elsewhere. Sausage is one of the goods to which it is said purchasers apply the price test for quality. It is difficult to believe that housewives are quite so gullible. Many are known to 'shop around' a practice which suggests that they prefer to test the quality of such things as sausage by flavour rather than price. Once they have found a brand to their taste, then high price or low, brand loyalty takes over.

There is more scope for a belief in the price test of quality in durable products where a real test of quality takes months or years in ordinary use. Purchasers cannot carry out the sophisticated tests used by the Consumers' Association before deciding to buy. It is in the field of new products such as clothing made of new man-made fibres and wares of new plastics materials where the problem of quality arises. There is no standard of comparison except the products of the older materials.

However, oddly enough, the classical examples of products which are said to have negative elasticities of demand are both non-durables: one is an alcoholic drink, and the other a car polish. The sales of both remained in the doldrums simply because, it seems, the price was too low to be believed. There have been cases of negative price-elasticities of demand for capital goods existing for a time in limited markets. One was the British machine tool, which enjoyed a negative price-elasticity of demand in the United States for a time after the Second World War.

Price experimentation is an empirical approach to pricing policy, and it must have been followed by those producers who have found that their products appear to have negative price-elasticities. There is always an element of doubt about results that do not conform to theory, especially theory that is confirmed by an overwhelming volume of practical experience. Such results should be treated with reserve. It is not usually possible to repeat pricing experiments on the same product in the same market more than once or twice in the course of two or three years. Repetition is often the only means of testing the force exerted by other factors which may be more potent than price change. New consumer goods have been known 'to take on' suddenly. If 'taking on' happened simultaneously with a price increase, it would look as though the price increase had caused the rise in sales. Hence the need for caution when drawing inferences about the influence that price change or any other particular factor has on sales.

However, the evidence both theoretical and empirical that some products have negative price-elasticity of demand is enough to make denial unreasonable. At the same time, it must be admitted that there is an upper limit to the range of prices at which the elasticity is negative. Where a producer continues to raise his price, there is bound to come a time when a further rise would result in a decrease in total sales by quantity, although his total sales-proceeds might record an increase. It is both interesting and instructive to build an imaginary sales schedule in which the price-elasticity of demand changes from negative to positive as the price rises. Table 8.5 contains such a schedule, which is a reproduction of Table 11.3 of my *Planning Profit Strategies*. The table shows

TABLE 8.5 An imaginary sales schedule

Price p £	*Quantity sold* Q *No.*	*Sales proceeds* $G=pQ$ £	*Price-elasticity of demand*
1	10	10	Negative
2	40	80	Negative
3	70	210	Negative
4	100	400	Negative
5	120	600	Negative
6	100	600	0·600
7	90	630	0·778
8	80	640	1·000
9	70	630	1·286
10	60	600	1·667
11	50	550	2·200
12	40	480	3·000

that the maximum quantity sold is reached just before the price elasticity changes from negative to positive, but that the sales proceeds continue increasing until they reach a peak when the elasticity of demand is unity. As the relationship assumed in the table between p and Q for prices with a positive elasticity is linear, the figures of the price-elasticity of demand in the last column are calculated by applying the formula of equation (xii) above to the price and quantity for:

$$Q=160-10p$$

which is the equation that satisfies the figures for $p=6$ and upward in the table.

BRANDING

There was a time during the years up to about 1920 when most groceries were sold unbranded. The grocer received granulated sugar in sacks from the refineries, and he weighed and packeted it while the customer stood waiting. He received tea in chests, and butter, margarine and lard in tubs. With these goods, the grocer decided what make or blend his customers should receive, and the competition amongst producers was almost perfect, and the price-elasticity of demand high. Jams and marmalades lent themselves to branding: they were sold in earthenware lb and 2 lb pots in those days, with a greaseproof paper lid tied down with fine white string. Oddly enough, too, a great deal of soap was branded. There was one toilet soap called Monkey Brand, which was advertised somewhat negatively by the slogan 'Won't wash clothes', which may explain why it ultimately disappeared from the shelves. Competition amongst producers for these branded goods was far from perfect, for branding introduced a monopoly element into the market. The makers of Monkey Brand held the monopoly of Monkey Brand, and consumers themselves decide the make and brand they will buy.

The advantages to producers of branding have long been known. But the cost of weighing and packaging by hand more than offset the additionl revenue to be achieved by monopoly pricing. When weighing and packaging machinery became available, it was at first limited to loose materials such as sugar and tea. Sugar refiners and tea blenders were the first producers to make use of the new technology. It was not until after the Second World War that the inventors produced a machine that could weigh and package such goods as butter and margarine. Then the branding of groceries became fairly general; and attempts are now being made without much success to brand meat, fruit and vegetables.

Cheese with its many named varieties appears to be unbrandable, though admittedly branding is so important that ways and means may be found of overcoming the difficulty of making a brand of, say, *Cheddar* that is slightly different from Cheddar in flavour or texture and which would be acceptable to the epicure. A brand of anything to be successful must be distinguishable from others, and it must have a good reputation, which may take years to acquire. New pharmaceuticals are always given brand names, and this is as it should be from the manufacturer's point of view; it puts them one step ahead of potential competitors, who may have no research costs to defray.

'TWOPENCE OFF'

Whether the current advertising device epitomised in the phrase 'twopence off' is regarded by consumers as a genuine price reduction or a salesman's gimmick is a matter of some controversy amongst marketing men. By *genuine price reduction* is meant the kind of reduction that would have the same effect on sales as a reduction advertised as such and not as twopence off. Would it be reasonable to measure the price-elasticity of demand for the brand in the area or market by comparing sales before and during the period that the 'special offer' operates? The short answer is, probably no. On the one hand, consumers loyal to the brand tend to stock up during the period of the offer; on the other, those who do not care for it or prefer another brand tend to be suspicious: 'there's a catch in it somewhere'; they may even believe that the quality of the special offer is better or worse than usual, better, to impress new consumers attracted by the offer, or worse, to clear an inferior batch produced by mistake.

An industrial statistician faced with sales statistics affected by gimmicky advertising needs to be something of a psychologist and a student of consumer behaviour. To understand his raw material and to make the best of it, perhaps he should turn field-worker for a time, and visit the shops and supermarkets in the area to question retailers and customers about their reaction to 'twopence off', free gifts so called, competitions and the rest. He may be surprised by some of the answers.

9

The production function

Theory has little to say on the matter of the form that the production model should take. It can be either linear or logarithmic, but not of quadratic or similar form. However, there is some empirical evidence, all pointing to the logarithmic form. A logarithmic production model is given in Chapter 5, equation (xvii), of which the linear form is:

$$\log Q = \log K + a \log F + b \log W + c \log E \qquad \ldots \text{(i)}$$

where Q represents the rate of production; F, the labour force; W, the average weight or size of the products; and E, the cumulative output of the products in each period or at each establishment. It is proposed in this chapter to discuss the general objects of deriving production functions, and then to consider each of the several factors in turn.

OBJECTS

Economists have been talking about production functions for many years. Many different forms have been offered, most of them of academic interest only. Our concern here is with a model of practical value in providing an analytical basis for solving some of the problems of business management. If the company reduces its output by x per cent, how much of its labour force should it be able to save? If it increases its output by y per cent, how much additional labour will it need? If the company changes the designs and specifications of its products, how will the change affect output for a given labour force? These are questions the answers to which can often be provided by the application of regression analysis to routine statistical data.

Production functions in this sense are concerned with physical entities only, never with inputs in terms of money. Nor are they concerned directly or indirectly with the analysis of labour time. There may be a case for accounting for labour in terms of the pay roll rather than in terms of men or man-hours; but pay roll introduces unnecessary complications largely as a result of changing wage rates, so that the case is not a very good one. It is physical input that counts in production, not money input. Admittedly, labour input has to be financed, and the pay roll may reflect differences in the quality of the labour employed from time to time and from place to place. There are ways and means of meeting the problem of quality differences.

MEASURING PRODUCTION

Trade practice and convention often determines the physical terms in which production is expressed. A production function relating to the whole company or a division of the company assumes that different products can be added together to give a total in each of a number of periods, and the question arises whether the conventional unit of quantity is homogeneous enough over the whole range. Can n cwt of 6-inch wire nails be added to m cwt of $1\frac{1}{2}$-inch ovals and q cwt of 3-inch cut nails to give $n+m+q$ cwt of nails without stretching the limits of what is reasonable? It looks as though some kind of production index should be used in cases like this.

The relationship between one product and another in respect of labour is a complex one. As money value does not provide a link between product and labour, a production index weighted for labour content in the base period instead of price should be considered. Official index numbers of quantity produced, exported or imported are all weighted for price or unit value; and undoubtedly this sets a precedent and a fashion in quantity index number making. However, a production index weighted for price does not seem to serve the need of the analyst seeking to make an index of productivity; indeed the appropriate production index for such a purpose seems to be one weighted for labour content; and it is suggested the same goes for a production index intended to provide the basic statistics of output for the purpose of deriving the production function. Such an index is described and demonstrated in Chapter 12.

As it stands, the model of equation (i) is designed for manufacturers producing goods of the same kind but different sizes involving assembly, where weight or some other measure of size can reasonably be used for expressing the total output. Kitchen cupboards, cars, tractors, compression ignition or spark ignition internal combustion engines, electric motors and air compressors are examples. The effect on the production–labour relationship of variations in average size is taken care of by the second independent variable, W, whose purpose is more fully discussed below. In applications where production index numbers are made to represent total output, the factor would probably be redundant.

LABOUR

It is suggested above that for most purposes, the best two measures of labour are man-hours and numbers employed. Since the latter is a point-of-time figure, it is not a simple concept in practice. If the basic period to be used for the analysis is the week, and the work-force is paid by the week, there is no difficulty, the number employed in each week remains constant during the week. But if the basic period is a quarter or a year, it may mean taking an average of the weekly figures covering each period. The average of the first and last weeks would not often be accurate enough, since the period may have a deep trough or a high peak in the middle.

Another point to be settled is whether the average should be the AM or the GM. The model is logarithmic in its linear form, so that in theory the average should be the GM, that is, the AM of the logarithms. However, there is a problem: if the labour force happened to fall to zero in any one or more weeks owing to holidays or a strike, the GM would be zero and since the logarithm of zero is minus infinity, and minus infinity plus any number or numbers short of positive infinity is similarly minus infinity, then the AM of the logarithms would be minus infinity of which the antilogarithm is zero. There is a choice of alternatives open to the analyst: one is to omit all weeks with zero labour and output from the analysis, and the other is to adopt the AM of the natural numbers. Since the GM appears to be preferable, the choice should fall to the former.

SCALE OF PRODUCTION

There are disturbing factors such as improvements in organisation and advancing technologies; but as there is usually no statistical measure of them they have to be left out of account or covered by a series representing time. In theory, often confirmed in practice, economies of scale are reflected in the relationship between production and labour. The value of a, the coefficient of log F, then, can be expected to exceed unity; the shape of the graph with production, Q, measured on the horizontal axis, and labour on the vertical, is like that of Fig. 8.6: it rises from the origin with a downward concavity. Where output is tending to exceed the capacity of the plant, the value of a would tend to fall below unity, when the shape of the graph would be similar to that of Fig. 8.5.

It became fashionable amongst practical statisticians about the time of the Second World War to express the relationship between production and labour in terms of what came to be known as *the scale of production law*. The advantage of this was that it enabled the non-mathematician to understand the practical significance of the derived value of a. If Q is doubled the labour required per unit of production would fall to Y per cent of the original requirement. This Y per cent is the percentage scale of production law. Conversion is simple enough. We have:

$$Q=K_2F^a$$

$$\therefore\ F=K_3Q^{1/a} \qquad \ldots (a)$$

Let Y represent the percentage law. Then when Q is doubled:

$$2YF/100=K_3Q^{1/a}2^{1/a} \qquad \ldots (b)$$

Dividing (b) by (a), we have:

$$Y/50=2^{1/a}$$

$$\therefore\ Y=50\times 2^{1/a} \qquad \ldots \text{(ii)}$$

or linearly:

$$\log Y=1{\cdot}6990+1/a.0{\cdot}3010 \qquad \ldots \text{(iii)}$$

It will be seen from both (ii) and (iii) that when $a=1$, $Y=100$, that is, the rate of production increases or decreases *pro rata* to labour; and that when $a>1$, $Y<100$, and when $a<1$, $Y>100$. In normal circumstances, when a factory is producing within the capacity of the plant, the value of a should exceed unity and that of Y fall short of 100. It is always useful for the analyst to have in front of him a short conversion table giving the value of Y for each of a range of values of a. Table 9.1 suggests how such a table could be set out, the underlying equation being (iii) above. It is exceedingly unlikely that the value of a could ever fall to 0·50, giving a scale of production law of 200 per cent. It would mean that to double the rate of output it would be necessary to quadruple the labour force. However, the figures of 0·50 and 200 per cent make a point: that the economies of scale can be negative in certain circumstances.

TABLE 9.1 A conversion table for the scale-of-production law

Value of a (1)	$\frac{1}{a}$ (2)	$\frac{1}{a}$ *0·3010* (3)	log $Y=$ (3)+*1·6990* (4)	$Y=$ *antilog. of* (4) % (5)
2·00	0·50	0·1505	1·8495	70·7
1·75	0·57	0·1716	1·8706	74·2
1·50	0·67	0·2017	1·9007	79·6
1·25	0·80	0·2408	1·9398	87·1
1·00	1·00	0·3010	2·000	100·0
0·75	1·33	0·4003	2·0993	125·7
0·50	2·00	0·6020	2·3010	200·0

WEIGHT OR SIZE

It is implied above that the average weight or size factor makes an index of production for the company's products unnecessary. In any event it can be introduced only where product homogeneity exists. If average weight is used to measure the factor, then total production also should be expressed in terms of weight. The general idea is that a heavy article is cheaper in labour per unit of weight than a light article. The same goes for volume and yardage: a quart bottle of beer or tin of paint takes less labour to produce and pack per pint than a pint bottle or tin. The weight or size factor takes care of this, and what is more it provides a measure of its effect on the production–labour relationship, and thus useful information for costing purposes.

Some practical statisticians like to express the labour per unit of output as a function of average weight and to call the resulting index of W the *average weight law*. Converted to a labour function, the production function of equation (xvii) in Chapter 5 reads:

$$F=K^{-1/a}P^{1/a}W^{-b/a}E^{-c/a} \qquad \ldots \text{(iv)}$$

Holding other factors constant, we have:

$$F=K_4W^{-b/a} \qquad \ldots \text{(v)}$$

Let N represent the number of units of output; and L, the labour per unit, then:

$$N = Q/W \text{ and } L = F/N = \frac{FW}{Q}$$

It follows that:

$$L = K_5 W^{1-(b/a)} \qquad \ldots \text{(vi)}$$

The average weight law is therefore $1-(b/a)$, which is of no help to non-mathematicians, but may assist the statistician in making comparisons and seeing how weight or size affects total output in relation to labour. I suspect the concept originated in the days before the Second World War when the practical value of regression analysis was little appreciated, and when statisticians attempted to derive the parameters of the production function piecemeal. In the USA, for instance, aircraft production statisticians[1] assumed that labour and weight bear the same relationship one to the other as the surface area of a cube bears to its weight, i.e.:

$$L = K_6 W^{2/3} \qquad \ldots \text{(vii)}$$

It does not take much imagination to realise that the formula assumes that labour is far more sensitive to differences in W than it really is. The parameter $1-(b/a)$ is much greater than 0·667. In the British aircraft industry it turned out to be about 0·90 according to the most reliable assessment[2], the value of b being equal to about one tenth of the value of a. Since the value of b must generally be less than unity, the shape of the graph of the relationship between production and average weight or size (for a given labour force) is similar to that of Fig. 8.5.

LEARNING

Learning theory has been the subject of some research in the USA. The theory is that the longer a man is engaged on the same task, the more efficient he becomes, and the greater his productivity. The importance of the concept in the context of industrial relations and product design policy can scarcely be exaggerated. To achieve the maximum productivity from learning, industrial relations should be

[1] *Vide* T. P. Wright, 'Factors affecting the Cost of Airplanes', *Journal of Aeronautical Science*, February 1936, and Report 17 of the Aircraft Resources Control Office of the USA, 30 July 1943.

[2] *Vide* E. J. Broster, 'Productivity in the Wartime Aircraft Industry', *Aircraft Engineering*, June 1957.

such that labour turnover is kept to a minimum, and changes in products and in the design of products should not go beyond the essential.

Needless to say, the effect on productivity of learning varies enormously from industry to industry: in machine-paced industries, it can scarcely count for anything at all; whereas in labour-intensive industries it can count for a great deal, although, if American research is anything to go by, rising productivity through learning is by no means restricted to labour-intensive industries.

The measure of learning is the cumulative output of products currently being produced. To avoid zero in respect of products first produced in the period, the cumulative output to the end of the period should be used. It is the average cumulative output of the several products in current production that is needed for the purpose, the appropriate average being the GM, which explains why zero is to be avoided.

It seems to be generally accepted that learning yields to the law of diminishing returns; that is, the parameter, c, is less than unity, and its graph is similar to that of Fig. 8.5.

A so-called percentage learning law has been expounded for the benefit of non-mathematicians and to assist statisticians to a clearer understanding of the phenomenon. It is similar to the scale-of-production law: each time the average cumulative output doubles, the labour required per unit of output falls to X per cent. This X per cent represents the learning law.

Holding other factors constant, we have from equation (iv) above:

$$F = K_7 E^{-c/a} \qquad \ldots (c)$$

When E is doubled:

$$XF/100 = K_7 (2E)^{-c/a} \qquad \ldots (d)$$

Dividing (d) by (c), we have:

$$X/100 = 2^{-c/a}$$

$$\therefore X = 100 \times 2^{-c/a} \qquad \ldots \text{(viii)}$$

A table on the lines of Table 9.1 showing how the values of X for a range of values of c and c/a can be compiled for ready reference. For the purpose, the parameter a can be held constant at, say, 1·0 or perhaps more realistically at 1·25, which is equivalent to a scale of production law of 87 per cent. Of course, if a good estimate of a is available, it should be preferred to any other figure. Table 9.2 shows the percentage learning laws for the kind of range of c and c/a that may be found in practice, though admittedly, percentage laws below 75 occur only rarely. For the purpose of the table, the value of a is assumed to be 1·25, the basic equation being:

$$\log X = 2 - c/1{\cdot}25 \times 0{\cdot}3010$$

Finally, a word about other potential factors may be worth while. The production function is concerned with the relationship between output and labour and factors that influence that relationship. Materials form no part of it except in so far as they change in a way that affects the extent of the processing necessary. A change from make to buy or from buy to make is an example. Here, there is a break in the relevant statistics, and unless they can be corrected realistically on one side of the break or the other, the only way round the difficulty thus created is to use the statistics on one side only, or on each side separately. An analysis on each side would have the advantage of giving an independent measure of the change in the output–labour relationship resulting from the change in make-or-buy policy.

TABLE 9.2 A conversion table for the learning law

Value of c (1)	$\frac{c}{1{\cdot}25}$ (2)	(2) × *0·301* (3)	log$X$= 2−(3) (4)	X= antilog. of (4) % (5)
0	0	0	2·0000	100·0
0·10	0·08	0·0241	1·9759	94·6
0·20	0·16	0·0482	1·9518	89·5
0·30	0·24	0·0722	1·9278	84·7
0·40	0·32	0·0963	1·9037	80·1
0·50	0·40	0·1204	1·8796	75·9
0·80	0·64	0·1926	1·8074	64·2
1·00	0·80	0·2408	1·7592	57·4
1·25	1·00	0·3010	1·6990	50·0

A more statistically amenable factor is the number of different types being produced, the greater the number, the smaller the rate of output in terms of tonnage or volume or of number of units for a given amount of labour. In my experience, the factor has proved to be of little if any significance; but that is probably because the difference in terms of labour per ton (or per gallon or per yard) between one type of the product and another was negligible. At any rate, there is no harm in taking the number of types in current production into account and testing the factor for significance. If the factor is not significant, there will be little if any improvement in the coefficient of multiple correlation or in the standard error of estimate; and the values of the parameters a, b and c will tend to lie out in the blue when the factor is adopted to serve as the dependent variable in an analysis by least squares. There is always a chance, too, that the non-significance of a potential factor will be discovered in the preliminary analysis by one of the less sophisticated methods of regression analysis. It is worth mentioning that where the dependent and independent variables are negatively correlated, as output is to the number of types in current production, the graph expressing the curvilinear relationship between them is similar to that in Fig. 8.3. A linear relationship of two factors negatively correlated takes the shape of the graph in Fig. 8.1.

10

Variable costing

Variable cost unqualified relates to that part of the total annual cost of production (or goods or services) which varies *pro rata* to the annual rate of output. Since the variable cost is part of the total annual cost, the total annual cost can also be said to vary with the rate of output, and to this extent, it too is a variable. However, the object of variable costing is to segregate those costs that vary directly with the rate of output from those that do not in order that the marginal cost may be determined, the marginal cost being defined as the net additional cost to the firm of increasing the rate of output by one unit; so that it is equal to the total variable cost divided by the rate of output, from which it follows that the total annual variable cost is equal to the marginal cost multiplied by the annual output.

This applies only to conditions in which the rate of output does not exceed the normal capacity of the plant. It also assumes that the marginal cost is the same for all rates of output within the normal capacity of the plant. *Normal capacity* can be defined in terms of the marginal cost: it is that rate of output beyond which the marginal cost tends to increase as a result of the special measures that have to be taken to increase output, such as the introduction of overtime for direct and indirect labour. It can also be defined in terms of that rate of output beyond which higher rates of wages have to be paid for direct labour and beyond which one or more of the non-variable items of annual cost tend to become variable. Figure 10.1 demonstrates the proposition graphically.

Just as the variable cost can be denoted as that part of the total cost that varies *pro rata* to output, so it is sometimes connoted as direct wages, materials, power, repairs (wholly or in part) to machinery and plant, royalties, and certain kinds of advertising. But as we have seen, this connotation is necessarily limited to rates of output within the normal capacity of the plant. It is also incomplete.

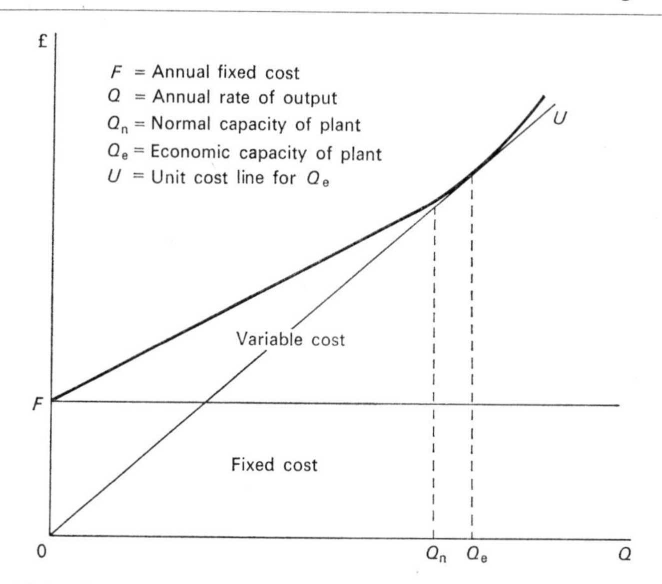

FIG. 10.1 Annual cost graph showing plant capacity

There may be objections to the assumption that the marginal cost remains constant for all rates of output within the normal capacity. The concept of the marginal cost is of academic origin, and it is still apparently supposed by academic economists that it falls as the rate of output rises from zero, reaches a level where it is constant, and then begins increasing as the output continues to rise. With output measured on the horizontal axis and the marginal cost on the vertical, the graph of the marginal cost would take the shape of the well-known U-curve. However, this academic idea does not conform to experience. In practice, the graph of the marginal cost is nearly always a horizontal straight line, and it does not tend to curl upward until the rate of output begins to exceed the normal capacity, and it rarely, if ever, has a downward slope over any range of outputs at all.

TIME COSTS

Fixed costs, so called, make up the remainder of annual costs. We shall be concerned with the *variable fixed costs*, which since the phrase sounds like a contradiction in terms, are best referred to as *time costs*: they vary *pro rata* to the length of the unit of time. A definition of time costs by connotation depends largely upon the type of business; but it would generally cover local rates, rent of land, buildings, machinery, plant and vehicles, some routine advertising, salaries, indirect labour, interest on loan capital, depreciation of wasting fixed assets, heating, lighting, office expenses, possibly a part of repair costs, and day-to-day maintenance.

TOTAL ANNUAL COSTS

Given the marginal cost and the total time costs, we can build up the total annual cost for any rate of output within the normal capacity of the plant. If the rate of output is 100,000 units a year, the marginal cost £2, and the annual time costs are £50,000, then the total annual cost is:

	£
Variable cost:	
100,000 units @ £2 each =	200,000
Time costs	50,000
Total annual cost	250,000

If we calculated the total annual cost for three other rates of output, and plotted these totals on graph paper with the quantity of output measured on the horizontal axis and the total cost on the vertical, the dots would describe a straight line which rises to the right, and which, if extended to the left, would cut the vertical axis at a point above the origin equal to the total time costs. That is, for zero output, the total annual cost is equal to the sum of the time costs (see Fig. 10.1).

This proposition is valid enough in theory. The difficulty is that it appears to be based on an inspection of individual items of cost and their dichotomous classification into variable costs and time costs. If a factory ceased production for a week or two, that is, reduced its rate of output to zero for the period, the company would be able to save not only its direct labour, but also most if not all its indirect labour. Salaries, too, could be cut down. Nevertheless, some time costs would continue to be incurred: rent, rates, interest, depreciation, and in some cases heating—what may be called the hard core of time costs.

It appears then that some of the time costs are at least semi-variable: for if it is possible to reduce indirect labour to near zero for zero output it would seem that it is equally possible to cut them to some extent when the rate of output is reduced by, say, 50 per cent, and conversely that they would have to be increased to some extent when output is, say, doubled. This does not affect our denoted definitions of *marginal cost* and *time cost*; it merely goes to show that fully connoted definitions are impracticable.

An industrial plant that is idle but which is all set to begin production at a given rate of output would be incurring all the time costs appropriate to that rate defined by connotation; it may also be incurring the wages cost of direct labour, depending upon whether it was set to begin production at a moment's notice or a longer period's notice. There is one item of variable cost it would not be incurring and that is materials. It would necessarily have stocks of materials sufficient to begin production, but it would not be consuming any. Materials, then, form the hard core of variable costs. And oddly enough, if by materials we mean goods of all kinds, raw, semi-finished and finished, then the term can be used in respect of wholesalers and retailers, when they form not only the hard core of variable costs but often also the only item of variable costs by connotation.

THE MARGINAL COST

If we cannot distinguish variable costs and time costs by connotation, then how can we determine the marginal cost? The answer is simple. There are in fact at least two methods, one arithmetical (the method of finite differences); and the other statistical (regression analysis by least squares).

To apply the method of differences, the first step is to obtain estimates of the total annual cost of production for three different rates of output. These estimates should be realistic relative one to another. Where a plant is given over to the production of a single homogeneous product, there is not much difficulty in estimating the total cost. But where there are two or more products, difficulties may arise owing to the arbitrary allocation of time costs. The hard-core of time costs could very well be omitted entirely and only variable costs included. Some other allocated costs such as that of top management could go along with the hard core costs. Indeed, it would be necessary to treat any potentially variable allocated time costs with extreme care. Differences in them, one rate of output compared with another, should be realistically due to differences in the rate of output and not to any cost accounting fiction, which may be valid enough for some other purposes.

Needless to say, the estimates should be built up on standard conditions of wage rates, materials prices and so on. Extraneous factors must not be allowed to influence the estimates and destroy their comparability.

COST ANALYSIS BY FINITE DIFFERENCES

Where it is definitely known that the marginal cost is a constant for all rates of output, only two sets of data are necessary, that is, two cost figures for two rates of output. The object of the third set of data is to provide a check on the linear assumption as it is called and also on the accuracy of the estimates. The method of differences is based on the simple and incontrovertible fact that provided the estimates are correctly made in accordance with sound principles, the difference in annual cost is due to the difference in the rate of output, this difference in cost being the variable element in the total. Table 10.1 demonstrates the method applied to a hypothetical example.

TABLE 10.1 Applying the method of finite differences

	Annual output		*Annual cost*		
Estimate (1)	*Total* '000 (2)	*Difference* '000 (3)	*Total* £'000 (4)	*Difference (variable cost)* £'000 (5)	*Marginal cost* (5)÷(3) £ (6)
A	10	—	60	—	—
B	20	10	90	30	3
C	30	10	120	30	3

The two sets of differences in columns (3) and (5) represent the increases from A to B and B to C.

Since the two marginal cost figures are the same, the linear assumption can be accepted as valid for rates of output within the normal capacity. The total variable cost for each rate of output is equal to the marginal cost multiplied by the annual output, thus for estimate A it is £3 × 10,000 = £30,000; for B, £60,000; and for C, £90,000. The pure time costs included in the estimates are therefore £30,000 for each rate of output, that is,

		£
for A	£60,000 – £30,000 =	30,000
for B	£90,000 – £60,000 =	30,000
for C	£120,000 – £90,000 =	30,000

This identity of the time element is to be expected. Indeed, if it had turned out otherwise, only an arithmetical error could account for it.

The unit cost, that is, the average cost per unit for each rate of output is not shown in the table. It will be seen that it falls as the rate of output increases: for A, it is £6·00; for B, £4·50; and for C, £4·00.

NORMAL AND ECONOMIC PLANT CAPACITIES

If, in the example of Table 10.1, the normal capacity of the plant is 30,000 units a year, then, by definition, any increase in the rate of output above this figure will have the effect of raising the marginal cost. As the rate of output is increased, the marginal cost will tend to continue rising. At the same time, the unit cost will continue falling until a point is reached where it may remain constant for a short range of outputs and then begin to rise. At this point, where the unit cost turns from a downward trend to an upward, it is equal to the marginal cost, which will have continued rising to meet the falling unit cost. For rates of output in excess of that corresponding to the equality of unit and marginal costs, both will continue rising, but the marginal cost will tend to rise more rapidly than the unit cost. This may seem impossible, but as Table 10.2 shows, it is perfectly feasible. See also Fig. 10.1.

The rate of output which gives the lowest unit cost is equivalent to the economic capacity of the plant, which can also be defined as that rate of output which equates the marginal cost to the unit cost. It can further be defined as that rate of output up to which it would pay in ordinary circumstances to use the existing capacity unextended, and above which it would pay to extend it.

Strictly, each marginal cost figure shown in column (6) of Table 10.2 is the average over the range of 1,000 units of output a year shown in column (3). In circumstances like these, each addition of one unit to the rate of output has its own unique marginal cost. The table does not present time series of observations nor does Fig. 10.1 present a time trend.

The analysis contained in Table 10.2 is somewhat academic. But it

TABLE 10.2 The economic capacity

Estimate (1)	*Annual Output* Total '000 (2)	Difference '000 (3)	*Annual Cost* Total £'000 (4)	Difference £'000 (5)	*Marginal cost* (5)÷(3) (6)	*Unit cost* (4)÷(2) (7)
C*	30	—	120·00	—	—	4·00
D	31	1	123·10	3·10	3·10	3·97
E	32	1	126·41	3·31	3·31	3·95
F	33	1	129·93	3·52	3·52	3·94
G	34	1	133·66	3·73	3·73	3·93
H	35	1	137·60	3·94	3·94	3·93

* Normal capacity. It will be seen the economic capacity is slightly less than 35,000 units a year.

serves its practical purpose of showing how the marginal and unit costs can converge rapidly and ultimately cross over, in a situation where the rate of output exceeds the normal capacity of the plant. The special measures that have to be taken in such a situation depend upon the circumstances. If one or two shifts are being worked, the operating time of machinery and plant is extended by putting the labour force on overtime, and paying the production men at overtime rates for overtime, and the maintenance staff at night rates. If a three-shift system is being worked, there is less scope for extending the operating time of machinery and plant. Week-end working has to be resorted to at the appropriate higher rates of wages, and it may even be necessary to pay the maintenance staff at Sunday rates for an appreciable proportion of their time. There would be a tendency for the whole establishment to become more labour intensive.

With a company engaged in a consumer-service industry, the scope for extending the operating time of machinery, plant and vehicles is negligible, and the economic capacity of the company's fixed assets may be much the same as the normal capacity. A local bus company cannot increase its output of services by employing additional men and introducing a three-shift system. The output of a passenger transport service is measured in terms of passenger-miles travelled and paid for, not of vehicle-miles or seat-miles. There is a useful business ratio in this respect, it is the ratio of passenger-miles to seat-miles, which is variously called the *load factor* or the *loading coefficient*.

VARIABLE TIME COSTS

In Chapter 5 reference is made to a certain common misconception which distinguishes the short-term and long-term marginal cost. Those who accept the distinction would classify the marginal cost as defined above as the short-term marginal cost. It will be observed, however, that subject to changes in wage rates and materials and power prices, this marginal cost could exist for a very long time, many years in fact; but it would not make it a long-term marginal cost in the sense in which the phrase is used. The two phrases, *short-term* and *long-term*, are misleading. They are drawn from the jargon of academic economics and do not imply a time element at all. An example will demonstrate the significance of the two phrases.

In the example of Table 10.2, we have the position where the economic capacity of the plant is, say, 35,000 units a year. To meet a growing demand for the product, suppose the firm decides to extend the normal capacity to 40,000 units a year, the outlay on the new work being £250,000, with additional annual capital charges and other annual time costs of £100,000. The long-term marginal cost is supposed to take account of these additional costs; but it is not at all clear how this can be logically argued. If the extension is brought into use when the original plant is working to its economic capacity of 35,000 units a year, then it could be argued that to produce an additional unit, the firm spends £100,000 in annual costs on an extension, so that the marginal cost of this unit of output amounts to £100,003.

This appears to be consistent with the shape of the total cost-curve, which apparently rises suddenly and vertically by £100,000 along the ordinate of the 35,000th unit of output. But this is quite illusory: immediately the extension is brought into use, the whole cost curve rises bodily over its entire length. It also extends further to the right by 10,000 units, from an original normal capacity of 30,000 units a year to a new one of 40,000 units.

It will be observed that the additional annual time costs of £100,000 are incurred, not in producing additional units of output, but in providing additional capacity. It is true that the extension has to be carried out before there can be any profitable increase in output beyond 35,000 units a year; but the original plant had to be provided before there could be any output at all. There is no logical reason why the annual time costs incurred in extending the plant should be treated differently from the annual time costs incurred in providing the original plant. They are both non-variable elements in any marginal costing system.

BASIC TERMINOLOGY

But they are non-variable only in the sense that they do not vary with changes in the rate of output. Relative to changes in the normal capacity of the plant, they are variable costs.

The opening sentence of this chapter refers to *variable cost* unqualified. In any context which deals with cost variability in general, it follows from the foregoing paragraphs that cost variability relating to the rate of output needs qualifying; and since this chapter is concerned with cost variability in general, costs that vary with the rate of output will be referred to as *product-variable costs*; costs that vary as a result of the introduction of a new capital work will be referred to as *project-variable costs*, and the time element included in the latter as *project-variable time costs*. The marginal cost will go unqualified; it will remain as defined above. And needless to say, the evidently illogical distinction between short-term and long-term marginal costs will be ignored. The excess of the so-called long-term cost over the short-term cost is taken care of in the project-variable time costs.

Whereas the product variable cost can be fairly precisely defined in any context, this does not hold true of the project-variable time cost. Admittedly, we can define precisely the total time cost of an undertaking or any part of it; but the project-variable cost depends entirely on the scope of the new work. This does not necessarily mean the size of the work or of the capital outlay. For instance, a factory extension may not call for additional stores keepers, handling labour, wages clerks and the like. In that case, stores and handling wages and wages clerks' salaries are non-variables. But if there is no land available for an extension and it is decided to erect a new factory away from the original premises, then additional stores keepers, handling labour and clerical staff would be required, and their wages and salaries would constitute a project-variable cost. Scope then, is not related to size, for the new factory may provide a smaller addition to capacity than an extension. Accounting, secretariat and statistical staffs and top management would probably not be affected at all by either an extension or a new factory, so that the salaries bill for them would be non-variable.

Although the popular conception, to be frequently found in both academic and technical literature as well as in general works, of an investment project or capital work is a new factory or factory extension, it is unlikely that either by number or capital expenditure, such capital projects or works constitute much more than 10 per cent of all projects or works. New factories and extensions form part of a class of gross capital formation, i.e. expansion-development, which in total probably constitutes no more than 20 per cent of the aggregate industrial capital expenditure in the United Kingdom.[1]

Of capital works there are six main classes which are relevant to variable costing, and these can be sub-divided by reference to the effect on the company's cost structure.

PROJECTS AND ASSETS

In the literature on discounted cash flow, a subject which is examined briefly in the next section, there is evidence of a great deal of confusion between the terms *project* and *asset.* Capital projects provide for the acquisition of assets of one or more different kinds: tangible or intangible, wasting or non-wasting, fixed or current, the last named taking the form of any necessary increment in the stocks of materials, work in progress, and finished goods. Only tangible wasting fixed assets are renewed or replaced, but one finds in the literature such phrases as 'replacement of the project', 'depreciation of project', which make sense only if one reads *asset* for *project.*

A project is a plan, an idea, something on paper. Apart from intangibles such as goodwill, assets are concrete entities which give a capital project a physical reality in the shape of a capital work. All wasting fixed assets including leases have finite lives, usually predeterminate with a fair degree of accuracy. Non-wasting assets such as land and earthworks have perpetual lives. Wasting assets call for annual provision for renewal or amortisation as a capital charge. Non-wasting assets do not. Projects, or rather capital works, generally have indefinite lives, which for all practical purposes mean perpetual lives, and this holds true whether the new works consist entirely of short-lived fixed assets or not. Asset life and project life are not in any way interdependent. Yet the literature contains scores of statements to the effect that the cash flow of net receipts be discounted over the life of the project, and nearly all examples demonstrating a particular discounting technique, ascribe a definite life to the hypothetical project.

[1] See E. J. Broster, 'An Economic Analysis of Fixed Investment', *The Economic Journal*, December 1958.

Some capital projects provide for one asset only, and where this is so not much harm can be done by supposing that the life of the project is equal to that of the asset; although where it is anticipated that the project will go on living far into the indefinite future beyond the life of the asset, the investment appraiser's assessment must understate the true profitability of the scheme by an appreciable margin.

The real difficulties lie with multi-asset projects. It rarely if ever happens that the several assets provided under a capital project have all the same life. Building work has a longer life than machinery and plant, mechanical machinery usually has a longer life than electrical machinery, and electrical machinery has a longer life than heavy road vehicles. Since the estimated gross revenues or net savings in annual costs from a multi-asset capital project are nearly always indivisible, that is, they cannot be allocated amongst the several assets, it is not possible to assess the anticipated financial results of each asset separately on the basis of the individual asset life.[1]

INVESTMENT CRITERIA

One of the problems that faced the early cost accountants given the task of appraising a capital project was how to bring the capital sum of the outlay and the periodic flow of operating costs to terms so that they could be added together without violating the rules of arithmetic. They had two ways of solving the problem, one a common-sense solution, and the other a nonsense solution. Both are in common use today—oddly enough, the latter probably more so than the former.

The common-sense solution

This converts the sum of the capital outlay to an annual cost, which can then be added to the annual operating costs incurred by the work. The total is then deducted from the average annual gross receipts expected to accrue to the projected new work. The conversion process spreads the original cost less residual value of each wasting fixed asset evenly over its life. Non-wasting assets such as freehold land, additions to materials, planning costs and the like are not depreciated, except that such items as planning costs would be amortised over the life of

[1] However, it can be done but only by first employing a technique that gives a correct assessment of the financial results by taking account of asset lives and not project life. See E. J. Broster, 'The Investment Criterion: Is Present Value Practicable?' *The Bankers' Magazine*, May 1964. The technique is the Annual Value (AV) which is discussed and demonstrated later in this chapter.

the work where it is limited. Leasehold land and building are amortised over the life of the lease.

An added refinement is the charging of interest on the total capital outlay, or on the total expenditure on non-wasting assets, and on half the expenditure on wasting assets on the assumption that on average over the lives of the wasting assets half the expenditure on them is recovered through depreciation provision. This last method was favoured by the Treasury until recently and has come to be known as the Treasury Method or TM.

The nonsense solution

This is the pay-back period, which may be good enough for a rough and ready check on a project, but which fails completely to give an estimate of the financial result. Pay-back operates in the opposite way to the common-sense method. It determines the number of years required to recover the capital outlay from the flow of net receipts, defined as the net saving in annual costs before interest and depreciation, or as the gross revenue less the total operating costs. In short, pay-back converts the yield before interest and depreciation to a capital sum by a straight year-by-year addition, and continues this process until the cumulative yield equals the outlay on the capital work. But the number of years it takes to do this is no measure of the profitability of the work: it is no more than a rough and ready measure of the break-even point. The profit necessarily lies in the years subsequent to the break-even point, and they are not taken into account at all.

Discounted cash flow

Since the Second World War supposedly more refined, sophisticated and rigorous techniques for bringing the capital sum of the outlay and the flow of net receipts to terms have been developed. They not only carry out a conversion process but also take care of the time factor reflected in the rate of interest.

There are five such techniques now generically referred to as Discounted Cash Flow or DCF. A brief definition and description of each of them is as follows:

1. *Internal Rate of Return* (*IRR*)—the rate of discount that converts the flow of net receipts over the life of the project to a capital sum equal to the outlay on the project. Where the project provides for a single wasting

asset, the discounting process can be performed over the life of the asset instead of the project. The main difficulty with this technique arises with multi-asset projects. Most capital projects have indefinite lives, and the net receipts from them are indivisible. IRR can meet this by using the life of the asset where only a single asset is provided by the project. But where there is a plurality of assets with different lives, IRR is impracticable: there is a cycle of renewals of assets that IRR cannot provide finance for. Another practical difficulty is that to determine the internal rate of return the time-consuming method of trial and error has to be used. There is also a fundamental logical objection to the technique. The internal rate of return is not a rate of interest, and as the master himself—the late Lord Keynes—has pointed out in the *General Theory of Employment, Interest and Money* (p. 137), the yield of an asset can be capitalised only with the appropriate rate of interest. It should be mentioned that the internal rate of return as defined above is identical to Keynes's marginal efficiency of capital, which he asserts is not a rate of interest.

2. *Net Present Value* (*NPV*)—the capital sum left after deducting the sum of the outlay from the present value of the net receipts discounted at the appropriate rate of interest over the life of the project. Here we have the life of the project again, so that this technique, although superior to IRR on logical grounds is open to the same strictures as IRR in respect of multi-asset projects. Otherwise NPV is simple to apply and does not require the use of the method of trial and error.

3. *Sinking Fund Rate* (*SFR*)—like IRR does not charge interest on capital but converts the capital outlay to an annual cost by charging depreciation on the wasting fixed assets on a sinking fund basis at the appropriate rate of interest. This technique is flexible and adaptable to all kinds of capital projects including multi-asset projects. Its drawback is that it does not charge interest on capital.

4. *Annual Value* (*AV*)—like NPV charges interest on capital. It converts the capital outlay to an annual value, hence its name, and it is precisely the same as SFR except that it charges interest on the outlay. This annual value is the annual capital charge, which is added to the annual operating costs or deducted from the annual net receipts. It should be mentioned that AV was in use long before the Second World War.

5. *Terminal Valuation* (*TV*)—strictly a compounding technique, not discounting. It compounds all expenditure (capital outlay and annual costs) and revenue at the appropriate rate of interest over the life of

the project, deducts compounded expenditure from compounded revenue to give a net capital value of the project at the end of its life. TV is a topsy-turvey version of NPV; and its solution yield is the NPV yield left to accumulate over the life of the project at compound interest, or conversely, the NPV yield is the present value of the TV yield. The amount of £ left to accumulate at 7 per cent compound interest for 20 years is £3·870, and the present value of £ due in 20 years at 7 per cent compound is £0·25842. These two figures are the reciprocal one of the other; if the NPV of a project with a life of 20 years is £1,000, its TV at 7 per cent compound would be:

£1,000 at £3·870 per £=£3,870

and the NPV of a project with a life of 20 years and a TV of £3,870 at 7 per cent would be:

£3,870 @ £0·25842 per £=£1,000

TV is open to the same strictures as NPV. It requires a terminal date in a world where prospective terminal dates are few and far between.

Of the five DCF techniques, there can be no doubt that AV is the best. As well as being logical, realistic, straightforward, practicable and adaptable, it has the added advantage over IRR, NPV and TV that it converts the capital outlay to an annual capital charge, so that it can be used in cost accounting generally and, more important, in cost analysis. As we have seen SFR is much the same as AV. It can be used by those analysts who prefer to calculate financial results gross of interest on capital.

The sinking fund

Since the sinking fund plays a central role in DCF techniques, it is as well that we should know how it works. Suppose a garage owner decides to run a car-hire service. He buys a car for £1,000, and he estimates its working life to be five years. At the end of this period, he will need to renew it, and his problem is to set aside each year a sum of money sufficient to provide for the renewal. He forecasts that owing to rising prices the replacement price will be £1,100, but he expects the residual value of the car will be £100, leaving £1,000 to be found in five years' time. He could, of course, divide £1,000 by 5 to give £200, which is the straight-line provision; but he decides to place his money in an investment which yields 5 per cent a year without risk of loss of capital, and to allow the interest to accumulate in the investment, so that each annual sum invested will accumulate at a compound rate of interest of 5 per cent.

Now, if he invested £200 a year at 5 per cent compound, it would amount to more than £1,000 at the end of five years. In fact, it would come to:

$$£200 \times £5{\cdot}526 \text{ per } £ = £1{,}105$$

The figure of £5·526 is the amount of £ a year accumulating at 5 per cent compound a year for five years. If £200 a year gives £1,105 at the end of five years, the annual amount to give £1,000 would be:

$$\frac{1{,}000}{1{,}105} \text{ of } £200 = £181{\cdot}0$$

but there is no need to go all round the Wrekin like this. Most published compound interest tables give sinking fund figures for a range of years and interest rates, and we can arrive at £181·0 by using such a table. The sinking fund figure is the annual sum required expressed as a decimal of £ to give £ at the end of the period. The annual figure to give £ at the end of five years at 5 per cent is £0·18097 (which is the reciprocal of £5·526 above), so that to provide £1,000 on the date required, we have:

$$£1{,}000 \ @ \ £0{\cdot}18097 \text{ per } £ = £181{\cdot}0$$

He therefore sets aside £181 a year as provision for renewal. The way it accumulates over the years throws further light on the sinking fund. Table 10.3 shows the year by year details for the hire-car example. There is, of course, no real need to make up such a table as this. If he wishes to know the amount standing in the fund at the end of some intervening year, all he need do is to calculate the amount of £181 a year for the appropriate year end. For instance, if he wishes to know the amount standing in the fund at the end of the third year, all he takes is the amount of £ a year at 5 per cent for three years from the appropriate table and multiply it by £181. Thus:

$$£181 \ @ \ £3{\cdot}152 \text{ per } £ = £570{\cdot}51$$

compared with the figure of £570·60 shown in Table 10.3. The differences here and in the figure of £1,000·14 in the Table relative to the figure of £1,000 required are due entirely to rounding off. The table shows that the interest accounts for £95 of the total of £1,000. Column (2) shows how compounding, that is, interest on interest, causes growth in the annual rate of increase: in the four years in which interest is credited to the fund, the rate of annual increase grows from £9·05 to £10·48.

TABLE 10.3 Sinking fund to give £1000 in five years at 5 per cent

Year	*Amount in fund at beginning of year down from col* (4) (1)	*Interest on amount in fund at 5%* (2)	*Annual provision* (3)	*Amount in fund at end of year* (1)+(2)+(3) (4)	*Annual provision plus interest* (2)+(3) (5)
1	—	—	181	181·00	181·00
2	181·00	9·05	181	371·05	190·05
3	371·05	18·55	181	570·60	199·55
4	570·60	28·53	181	780·13	209·53
5	780·13	39·01	181	1,000·14	220·01
TOTAL		95·14	905		1,000·14

A DEMONSTRATION OF THE ANNUAL VALUE METHOD

Suppose a chemicals manufacturer decides to extend the company's plant at a capital cost of £10,000, the following being the details of the additional assets:

	Life years	*Original cost* £	*Net replacement cost* £
Freehold land	—	1,000	—
Building work	50	2,000	1,200
Plant	20	3,000	3,000
Machinery	10	2,500	3,000
Works trucks	5	1,000	1,000
Additional stock of materials	—	500	—
Total cost:		£10,000	

The risk-free rate of interest appropriate to the sinking fund provisions is, say, 7 per cent, and the Company decides to charge interest on capital at its borrowing rate on unsecured loans, i.e., say, 10 per cent. Then by AV, we arrive at the annual capital charge as follows:

	£
Interest—	
£10,000 at 10%	1,000
Renewal provision—SF @ 7%—	
Building: 50 years—	
£1200 @ £0·00246 per £ =	3
Plant: 20 years—	
£3000 @ £0·02439 per £ =	73
Machinery: 10 years—	
£3000 @ £0·07238 per £ =	217
Works trucks: 5 years—	
£1000 @ £0·17389 per £ =	174
Total annual capital charge:	1,467

It will be observed that interest is charged on the total outlay, but renewal provision on the net replacement cost of the wasting fixed assets only. Land is not as a rule a wasting asset, and provided it is not subject to erosion by the sea or to inundation by sand or floodwater, its value is likely to appreciate rather than depreciate. When the total annual capital charge has been calculated, it is added to the other annual costs or deducted from the annual net receipts.

One of the advantages of AV over IRR and NPV is that it takes into account the differences in characteristics of the several fixed assets that ultimately convert a capital project into a physical reality.

So far, the question of discounting the cash flow of net receipts has not arisen. Indeed, it is a problem only where the cash flow varies from year to year, provided AV is used. If IRR or NPV is used, the cash flow is discounted whether the cash flow is constant or a variable. The discounting process is perfectly simple. For AV all that is required is the discounted annual average of the net receipts before the annual capital charge is deducted. First it is necessary to turn up the table of the present value of £ for each of the first few years at the risk-free rate of interest. Usually, where the life extends beyond five years or so, only the first few years' estimated net receipts vary. We will return to this problem in a moment. Table 10.4 demonstrates the method of calculating the discounted annual average. The table contains three examples each discounted at 5 per cent compound: the first series of cash flow has an upward trend, the second is constant, and the third has a downward trend. It will be observed that the discounted annual average of the annual constant is equal to the annual constant itself.

The weighted averages are equal to the totals of columns (3), (5) and (7) divided by the total of the weights, i.e. 4·3294. This figure of £4·3294 is equal to the present value of £ a year for five years at 5 per cent. As the five-year annuity which a £ invested at 5 per cent would purchase is the reciprocal of this, that is, £0·2310, the weighted averages are equal to the five-year annuities which the present values of the cash flows would purchase at 5 per cent; in short, the weighted averages are discounted averages.

TABLE 10.4 Discounted annual averages at 5 per cent

		First example		*Second example*		*Third example*	
Year	*Present value of £ at end of year* (1) £	*Net receipts* (2) £	*Present values* (2) × (1) (3) £	*Net receipts* (4) £	*Present values* (4) × (1) (5) £	*Net receipts* (6) £	*Present values* (6) × (1) (7) £
1	0·9524	1	0·9524	3	2·8572	5	4·7620
2	0·9070	2	1·8140	3	2·7210	4	3·6280
3	0·8638	3	2·5914	3	2·5914	3	2·5914
4	0·8227	4	3·2908	3	2·4681	2	1·6454
5	0·7835	5	3·9175	3	2·3505	1	0·7835
TOTAL	4·3294	15	12·5661	15	12·9882	15	13·4103
Averages—							
Unweighted		*3*		*3*		*3*	
Weighted		*2·9025*		*3·0000*		*3·0975*	

It is worth mentioning that the discounted annual average is a straightforward weighted average. It has been argued that the process is cumbersome, time-consuming and complicated; but this must be the result of a misunderstanding. Discounting, it should be kept in mind, is designed to take account of the interest value of money, that £ now is worth more than £ tomorrow. For discounting cash flows, the appropriate rate of interest to use is the risk-free long-term rate, which is equal to the current redemption yield on long-dated Government bonds.

Table 10.4 demonstrates the process of discounting for a project with a five-year life. Where the life extends beyond five years and the cash flow becomes a constant, the same method demonstrated in the table could be used. But where the life is very long or indefinite, a short-cut is available. In this the present value of an annuity or perpetuity deferred, say, five years is added to the present value of the variables for the first five years, the total of the weights adjusted accordingly, and the discounted annual average then calculated as shown in the table.[1]

COST ANALYSIS BY LEAST SQUARES

The foregoing excursion into the sphere of project costing is designed to assist the cost analyst towards a clear understanding of the nature of cost variability. Three cost functions were suggested in Chapter 5, viz.: equations (xiii), (xiv) and (xv). Of these, the first two are concerned with product cost variability only, and the third with product and capacity cost variability.

Least squares is a method that is scarcely worth considering unless the costs and outputs to which it is to be applied are actual as distinct from synthetic. Synthetic costs are costs built up by a cost accountant for a range of outputs for analysis by finite differences, the assumption being that the figures are mathematically accurate in the sense that any sets of synthetic figures in excess of the bare number required for solution by simultaneous equations would be perfectly correlated, and would therefore be worthless as degrees of freedom.

A cost analyst who chooses to reject synthetic costs as unrealistic must face many problems if the alternative is to apply a method of regression analysis to actual data in the form of time series. First, he must either render his cost series comparable by correcting the figures for changes in wage rates and materials prices or introduce them into his cost model as secondary factors, which would generally take the statistical form of index numbers. The latter can be rejected out of hand as likely to cause spurious correlation.

Another difficulty for the multi-product factory or company is the measure of output. For two or three products, there may not be much of a problem, except where the data available do not provide an adequate number of degrees of freedom. Then an equation of the form of the second of the three (xiv) could be used provided the series of costs are corrected as suggested above. With a multiplicity of products, an equa-

[1] For a detailed discussion of compounding and discounting including deferred annuities and perpetuities, see E. J. Broster, *Appraising Capital Work*, Longmans, 1968, Chapter 4.

tion of the form of the first one (xiii) could be used, output being expressed in terms of index numbers.

Two questions arise on this latter suggestion:

(*a*) What weighting system should be used for the production index?
(*b*) Would it be possible to calculate from the coefficient of the index, the coefficient i.e. the marginal cost, of each individual product?

We attempt to answer these questions in the following paragraphs.

The weighting system

There is a choice of two weighting systems, viz.:

1. prices or values in the base period, which is the system used in the making of official index numbers of production;
2. direct labour content in the base period.

Both systems are discussed in Chapter 12. For the purposes of cost analysis, since labour is more closely related to cost than price or value, it seems that a labour-weighted index is preferable to a price- or value-weighted one. However, it is a matter for consideration. There is a clear-cut case for preferring a labour-weighted production index for some purposes such as a necessary stage in making a productivity index, but it is not so clear-cut for the purpose of cost analysis.

The marginal cost of product

It seems that the coefficient of an index of production would be of little practical value except as a means of calculating the marginal cost of each individual product. The coefficient of the index is a marginal cost: it is the variable cost per unit of index. To determine the several product marginal costs, it would first be necessary to determine how many additional units of a product would have the effect of increasing the index by one unit (or by one point to use the jargon of index numbers). The appropriate figures to use are the base period weights and quantities. Table 10.5 demonstrates a method.

It is supposed that the coefficient of the index of output turns out to be 50, i.e., say, £50, and that there are four products, lettered A to D. The logic of the method is as follows: The total of column (3) is equated to 100 in the course of index number making; and we need to know by what quantity the value of V for each product in turn would have to

TABLE 10.5 Determining the product marginal cost where production index numbers are used

Product	*Weight* W (1)	*Output* Q (2)	$WQ=V$ (3)	$\frac{\Sigma V}{100W}$ (4)	*Marginal cost of product* £50 ÷ (4) (5)	*Check—* (2) × (5) (6)
A	10	100	1,000	16·500	3·030	303·0
B	25	300	7,500	6·600	7·576	2,272·8
C	20	200	4,000	8·250	6·061	1,212·2
D	40	100	4,000	4·125	12·121	1,212·2
			16,500			5,000·2

increase to give an increase in the index of one point to 101; for such an increase $\sum V$ would rise to

$$\frac{16{,}500 \times 101}{100} = 16{,}665$$

that is, by 0·01 of 16,500 = 165, which, divided by W, gives the increase in the output needed to raise the index by one point. As each point of the index has a variable cost of £50, the marginal cost of any given product is £50 divided by this increase in output. The marginal costs are shown in column (5) of the table.

A check on the method and calculations can be made by charging each value of Q in column (2) by its corresponding marginal cost to give the total variable cost, which should be equal to 100 × £50 = £5,000. The last column contains the check. Rounding accounts for the slight difference of £0·20.

It is questionable whether the capacity marginal cost could be calculated by regression analysis on the basis of the model of equation (xv) of Chapter 5. In any event, it would probably be much easier and more satisfactory to build it up on the basis of the annual capital charges of each unit of capacity particular to each product. It would be necessary to correct the value of the unattached constant, which would cover *inter-alia* the annual capital charges incurred in providing common-user fixed assets, that is, assets not attributable to individual products. The analysis of time costs between attributable and non-attributable costs is considered under Attribution Costing in Chapter 19.

It is worth mentioning that attribution costing, which has its mathematical expression in equation (xv) of Chapter 5, is a relatively new concept. There is probably much that has yet to be said about it in theory, and once the Institute of Cost and Works Accountants has accepted it as an important advance in the field of costing, much will be said about it in practice too, no doubt.

An entirely new technique in the same field is that of employing an index of output in a multi-product factory or company for use in the regression analysis of total annual costs, and deriving the several marginal costs in the way suggested above. Problems would undoubtedly arise in firms which produce and sell a wide variety of spare parts; and the statistician would have an opportunity of exercising his ingenuity to solve them.

11

Rational pricing

Pricing policy is the fundamental problem of business finance and management economics. A rational price is the price (or narrow range of prices) that gives the maximum net profit. For this reason, it is often called the optimum price; and any other price above or below is called a sub-optimum price.

In spite of the efforts of the schools of business studies and such writers as myself in preaching the gospel of rational pricing, price-fixing policy in industry still, it seems, plays but a minor role in the marketing of goods and services. Marketing has made rapid strides in recent years in two respects, viz.: in the techniques of activity and research, and in its position in the management hierarchy. There was a time not so many years ago when the production department decided what goods should be made and what services provided, and the marketing or sales department was expected to sell them. Now, according to the tenets of modern management philosophy, the marketing department acts like a customer of the production department, and orders the goods or services of predetermined design and specification in the quantities which it believes it can sell at a profitable price.

MARKETING AND PRICING

That is as it should be. Sophisticated though they are, however, modern marketing objectives appear to leave little room for rational pricing. The promotion of snob values, and brand loyalty, and maximising the company's share of the market can be, and indeed are, carried on without a thought to profit. An occasional case of alleged negative price-elasticity of demand tends to render the basic argument of rational pricing somewhat unconvincing. Many firms find themselves in a whirligig of short-term considerations with their accompanying gimmickry. It may be gratifying to top management to find that the company's share of the market is increasing. But is it profitable? Gratification is all very well; it does not pay the bills nor create profits for the shareholders. The criterion of efficiency is profit not market share. A sub-optimum pricing policy designed to increase market share is a policy of profit reduction.

MARGINAL REVENUE

It is true to say of any brand of a manufactured product that a reduction in price has the effect of increasing the sales proceeds. It also necessitates an increase in output and therefore in annual costs. When the increase in revenue or sales proceeds resulting from an increase of one unit in the rate of sales is equal to the marginal cost, the price is at its optimum. Such an increase in revenue is called the marginal revenue. When the price elasticity of demand is above unity, the marginal revenue is always positive; when it is equal to unity, the marginal revenue is zero; and when it is below unity, the marginal revenue is always negative. From Table 8.5, which demonstrates these propositions, it will be seen that as the price falls from £12 to £10, the quantity sold increases from 40 units to 60 units, and the revenue from 480 to 600. The marginal revenue for the intermediate price of £11 is equal to

$$\frac{600-480}{60-40} = £6$$

For a price of £10 it is £4; for £9 it is £1, and for £8 where the elasticity of demand is unity, it is zero. As shown later, the marginal revenue is equal to

$$p\left(1-\frac{1}{e}\right)$$

where p is the price and e is the elasticity of demand. For $p = 11$ and $e = 2{\cdot}200$, this would give:

$$11 \times \left(1-\frac{1}{2{\cdot}2}\right) = £6$$

Graphically, the measure of the marginal revenue is the slope of the annual gross revenue curve, and that of the marginal cost is the slope of the annual cost curve. Where marginal cost equals marginal revenue, the two curves are parallel. It is clearly shown in Fig. 5.4

that this corresponds to the maximum net profit, which is measured on the vertical axis.

BRANDING

The concept of optimum price is of academic origin; but it is none the worse for that. Academic economists were the first to appreciate the effect on the market mechanism of branding. Branding introduces a monopoly situation into the competitive field where high price-elasticities of demand for individual suppliers' goods reigned supreme. Now, with branding, each supplier holds the monopoly of his own brands, with the result that the price-elasticity of demand is relatively much lower, and the supplier has a freedom of movement in his pricing policy whereas previously he had no freedom of movement at all. Then, price was the dependent variable and supply and demand were the independent variables. Now demand is the dependent variable and price and supply are the independent variables; though admittedly there always was and still is a degree of interdependence amongst the three factors.

Branding is of business origin; but its effect on the market exercised the minds of academic economists for many years following the First World War. Undoubtedly, the business community exploited branding in its own interests; but economists thought businessmen had a greater understanding of the implications of branding than they really had. The result was that when they discovered the optimum price formula they ascribed it by implication to those firms which were branding their products. In fact, economists asked themselves the wrong question: 'How do businessmen exploit branding?' Whereas what they should have asked is, 'How could businessmen exploit branding to the best advantage?' In any event, it was the latter of these two questions that they answered.

MARGINAL AND AGGREGATIVE ANALYSIS

There are two distinct approaches to the rational pricing formula. One is that of marginal analysis, first propounded by academic economists in the early nineteen-thirties, and the other that of aggregative analysis which I first propounded in 1938 in an article in *The Accountant* and in my *Cost, Demand and Net Revenue Analysis* (Gee, 1938). I prefer the aggregative approach, partly because it is my own brain child, partly because it lends itself more readily to the analytical method,[1] but mainly because I discovered it in an attempt to solve a practical problem.

The origin of the aggregative method

It may be worth while my narrating how my discovery came about, if only to show that the optimum price formula is of more than academic interest, and that it has been applied satisfactorily to the solution of at least one practical problem of business importance. At that time in the nineteen-thirties, I was employed at Euston working for J. C. Stamp (later Lord Stamp) and filling in my time on capital-works efficiency auditing. Sometimes Stamp asked me to go along to his room to discuss the problem he had and to help him find a solution. At other times, when he was not in a great hurry for an answer, he would send me a note setting out his requirements. On the occasion in question, he sent me a copy of a report signed by the Chief Commercial Manager and designed to show the beneficial effects on passenger receipts of the appreciable reductions in fares which had been introduced. Stamp was too much of an economist to accept the report at its face value. Accompanying the report was a note addressed to me. It read as follows:

> *Mr Broster*
>
> Will you consider this for me in relation to changing means, i.e., to comment on the possible criticism that the change in trend of receipts is not due to cheaper tickets at all, but would have happened in any case, with changing prosperity, etc. ?
>
> I suggest, at a venture, ascertain the receipts curve 1931/6 on the basis of a ratio to income or prosperity after modifying the ratio by any annual factor representing loss of railway travel habit established during the previous rise of income (?1925–9) and then comparing that adjusted curve with the new actual.
>
> J.C.S.
> 16.4. (1937)

Stamp was so interested in my detailed reply, which confirmed his doubts only in part, that he suggested I should write a paper on the subject covering the four main-line companies, with a view to its

[1] The only complete marginal solution I am aware of is that given in Joan Robinson's *The Economics of Imperfect Competition*, in which the geometric method is used.

publication in one of the academic journals devoted to economics. The paper duly appeared under the title 'Railway Passenger Receipts and Fares Policy', in *The Economic Journal* of September 1937.

Importance of the marginal cost

However, I was far from satisfied. We all seemed to be regarding any increase in receipts resulting from the issue of cheap tickets as additional net profit. There had necessarily been an increase in the volume of passenger traffic and this could not have been carried without additional cost to the railways. At that time, although we understood and applied the principles of variable costing to capital works appraisal, the concept of the marginal cost was little known anywhere outside academic circles. What we needed was a cost equation expressing railway operating costs as a function of passenger-miles and net ton-miles. This would give the average marginal cost per passenger mile which I could compare with the average receipt per passenger-mile in each year.

I had assumed a constant price-elasticity of demand in my analysis of passenger receipts, and I now plotted the graph of the resulting equation on rectilinear graph paper. The graph described a curve rising to the right and concave downward like the revenue graph in Fig. 5.4. For the cost equation, I now assumed a constant marginal cost, which meant that the graph expressing cost as a function of passenger-miles would describe a straight line. Multiple regression analysis was applied to the available statistics of total operating costs of the four main-line railways, passenger-miles, which had to be partly estimated,[1] and net ton-miles. It was also applied to cost, coaching train-miles and loaded freight train-miles. The two analyses gave the marginal cost per passenger-mile, per net ton-mile, per coaching train-mile and per loaded freight train-mile.

It was at this stage that I began applying my mind to the problem of the optimum fare. I drew some sketch diagrams, showing the relationship between the annual revenue and annual cost curves for passenger traffic, amongst them one on the lines of Fig. 5.4. It became clear that a

[1] Passenger-mile statistics had been compiled for each of only four months in 1924 and 1925. Estimates could be arrived at in any of three ways: (*a*) dividing total passenger receipts by an estimated average receipt per passenger mile; (*b*) multiplying the number of passenger train-miles by an estimate of the average train load; and (*c*) multiplying the number of journeys by an estimate of the average length of journey. So far as I know British Rail uses one of these ways of estimating passenger-miles.

net revenue function could be derived by deducting the total cost function from the sales function, or a contribution function by deduction of the variable cost from the sales function. Since the time cost F could not be allocated between passenger and freight traffic, only the latter, the contribution function could be considered. Since the total cost graph and the variable cost graph, which begins at the origin, are parallel, the omission of the fixed cost could not affect the result.

Equation (viii) of Chapter 8 expresses the sales proceeds, S, as a function of price, p; the quantity of sales, Q; and the price-elasticity of demand as a constant, i.e.:

$$S = pQ = KQ^{1-1/e} \qquad \text{. . . (i)}$$

The corresponding annual variable cost equation is:

$$V = aQ \qquad \text{. . . (ii)}$$

where V is the variable cost; and a, the marginal cost. The contribution, which is represented by C, may be written:

$$C = S - V = KQ^{(1-1/e)} - aQ \qquad \text{. . . (iii)}$$

At its peak, the differential coefficient of C with respect to Q, is zero, i.e.:

$$\frac{dC}{dQ} = K(1-1/e)Q^{-1/e} - a = 0$$

$$\therefore \quad K(1-1/e)Q^{-1/e} = a$$

$$\text{i.e. } Q^{1/e} = \frac{(1-1/e)K}{a}$$

$$\therefore \ Q = \left(\frac{(1-1/e)K}{a}\right)^{e} \qquad \text{. . . (iv)}$$

This represents the quantity Q_m that would be sold at the optimum price, p_m. We can substitute it for Q in equation (vii) of Chapter 8, which reads:

$$p = KQ^{-1/e} \qquad \text{. . . (v)}$$

$$\text{i.e. } p_m = K\left(\frac{(1-1/e)K}{a}\right)^{-1}$$

$$= K\frac{a}{(1-1/e)K}$$

$$= \frac{ae}{e-1} \qquad \text{. . . (vi)}$$

which is the optimum price formula. Using a to represent the marginal cost per passenger-mile, I applied the formula to the passenger traffic of the four main-line railways and came to the conclusion that the companies had reduced their fares on average to a level somewhat below the optimum. The results of the regression analyses and the conclusions in respect of fares policy were published in a paper on 'The Variability of Railway Operating Costs' in *The Economic Journal* of December 1938.[1]

Since the optimum price is achieved when the marginal revenue is equal to the marginal cost, another approach to the formula is to equate the two values. From equation (i) above, we have the marginal revenue, M:

$$M=\frac{dS}{dQ}=(1-1/e)KQ^{1/e} \qquad \text{. . . (vii)}$$

Since $KQ^{-1/e}=p$, we can rewrite (vii):

$$M=(1-1/e)\,p \qquad \text{. . . (viii)}$$

For the optimum price then, we have:

$$a=(1-1/e)\,p_m$$

$$\therefore\ p_m=a\frac{e}{e-1}=\frac{ae}{e-1}$$

which is the same as equation (vi).

Since $K=k^{1/e}$, equation (iv) can be rewritten:

$$Q_m=k\left(\frac{ae}{e-1}\right)^{-e} \qquad \text{. . . (ix)}$$

Multiplying this by the optimum price, we have the optimum sales proceeds, S_m;

$$S_m=k\left(\frac{ae}{e-1}\right)^{-e}\left(\frac{ae}{e-1}\right)$$

$$=k\left(\frac{ae}{e-1}\right)^{1-e} \qquad \text{. . . (x)}$$

[1] At that time J. M. Keynes (later Lord Keynes) was the Editor of *The Economic Journal*. It was through his interest and encouragement that I came to write the paper at all in the face of much opposition at the office at Euston. Although Stamp himself was deeply interested in the analyses and conclusions, he wisely adopted a neutral attitude in respect of the opposition.

Another potentially useful equation can be derived directly from (viii) above:

$$M = p - p/e$$

$$p/e = p - M$$

$$\frac{1}{e} = \frac{p - M}{p}$$

$$\therefore\ e = \frac{p}{p - M} \qquad \ldots \text{(xi)}$$

As stated previously, the price-elasticity of demand changes only very slightly with changes in prices, so slightly indeed, that within any range of prices normally found in practice, it can be regarded as a constant. However, where the calculated optimum price lies outside the observed price range, a gradual approach to it is advisable. Where market research is carried out and the effect of the main disturbing factors can be eliminated, the management statistician should find equation (xi) particularly useful. The price-elasticity of demand tends to be higher for high prices than for low prices, so that if the calculated optimum price lies above those of the observed range, the statistician should expect if anything to find a higher price-elasticity of demand rather than a lower one for prices approaching the calculated optimum.

In using equation (xi) for such a purpose, the management statistician should keep in mind that the higher the price-elasticity of demand, the lower the optimum price. For it means that if the price-elasticity of demand rises significantly when price is increased, the originally calculated optimum price will be above the current true optimum price. An examination of the characteristics of equation (xi) is worth while. Under conditions of a constant price-elasticity of demand, the marginal revenue would rise or fall *pro rata* to price. Where the price-elasticity of demand rises with an increase in price, the marginal revenue would increase more than *pro rata* to price. If $p=4$ and $M=2$, then $e=2$, and if $p=6$ and $M=3$, then e remains at 2. But if when $p=6$, $M=4$, then e would be 3.

Equation (viii) could not be used for determining the marginal revenue for such a purpose—it would merely beg the question. It would have to be calculated from research statistics in the way described and demonstrated in the section above headed 'Marginal Revenue'. The appropriate formula is:

$$M = \frac{\Delta S}{\Delta Q} \qquad \ldots \text{(xii)}$$

where S is the sales proceeds in unit time; Q, the corresponding number of units of the product sold; and Δ denotes an increase or decrease in respect of a small change in price. The value of M thus obtained applies near enough to the arithmetic mean of the two prices. It is most important that the analyst should keep this in mind when attempting to estimate e by using the formula of equation (xi); for where e is a constant, so also is $(1-1/e)$, so that, as equation (viii) shows, M would vary directly and *pro rata* to price. The smaller the relative change in price, the more accurate is the calculated value of M.

A simple example will demonstrate the point and relate the method of approximation to the accurate method. Suppose it is found that the value of Q when p is 6 is 400, and when 8, Q is 225. Then by the accurate method, $-e$ is equal to the increase in the logarithms of Q divided by the increase in the logarithms of p, i.e.:

$$-e=\frac{2{\cdot}3522-2{\cdot}6021}{0{\cdot}9031-0{\cdot}1249}=\frac{-0{\cdot}2499}{0{\cdot}1249}=-2{\cdot}0$$

$$\therefore\ e=2{\cdot}0$$

Then for $p=6$, $M=\frac{1}{2}6=3$
for $p=8$, $M=\frac{1}{2}8=4$
and for the average price 7, $M=3{\cdot}5$.
It will be seen that the corresponding values of S for the example are 2,400 and 1,800 respectively. Applying the formula of equation (xii), we have:

$$M=\frac{600}{175}=3{\cdot}4286$$

compared with 3·5. Solving for e by applying the formula of equation (xi) to $p=7$ and $M=3{\cdot}4286$, we have:

$$e=\frac{7}{7-3{\cdot}4286}=\frac{7}{3{\cdot}5714}=1{\cdot}9600$$

Although the change in price from 6 to 8 is not by any means insignificant, the errors in the calculated values of M and e are quite small. If the price had changed from 6 to 7, then for $p = 7$, $Q \doteqdot 294$ and $S = pQ = 2{,}058$, and for the average of the two prices, 6·5, we have:

$$M=\frac{2{,}400-2{,}058}{400-294}=\frac{342}{106}=3{\cdot}227$$

compared with $\frac{1}{2}6{\cdot}5=3{\cdot}25$. Applying the formula of equation (xi), we have:

$$e=\frac{6\cdot 5}{3\cdot 273}=1\cdot 986,$$

which is a much closer approximation to the correct value of e than that obtained above.

Limitations of the formula

Econometric formulae always have their limitations; the optimum price formula is no exception. Where the price-elasticity of demand does not exceed unity, the answer it gives makes nonsense. For $e=1$, the optimum price appears as infinity; for $e<1$, it appears as some negative value. Oddly enough, the smaller the value of e, the greater the value of $e/(e-1)$, so long as $e>1$. For $e=2$, $e/(e-1)=2$; for $e=1\cdot 5$, $e/(e-1)=3$; for $e=1\cdot 1$, $e/(e-1)=11$; and for $e=1\cdot 001$, $e/(e-1)=1{,}001$.

Where the price elasticity of demand for a brand stands at unity or below—a situation which would rarely if ever be found in practice —the selling price can be said to be too low. An increase in price would reduce the quantity sold and therefore the cost of production, and if $e<1$ it would have the effect of increasing the sales proceeds. This process of increasing the selling price could be continued until it began to influence the price elasticity of demand, and raise it above unity. Small increases in price would probably have no effect. The pricing policy maker must think in terms of increases of 50 per cent and more. Whatever is done, it should be done in consultation with the management statistician and the market research office.

At the other end of the scale, there is no limit, in theory at any rate. The value of the expression $e/(e-1)$ could approach unity without ever reaching it, the higher the value of e. This means that the optimum price would approach the marginal cost, also without ever quite reaching it. However, in practice, it is unlikely that the price-elasticity of demand for a brand of any product could ever exceed, say, 10, which would mean that the competition in the market is so fierce that some companies would be compelled to withdraw and so ease the situation for other companies' brands. A high brand-price-elasticity of demand thus tends to be self-correcting.

It is important to distinguish between the brand-price-elasticity of demand and the product-price-elasticity of demand. The product-price-elasticity of demand may be very low, certainly much lower than the brand-price-elasticity of demand. For products in popular demand such as bread, potatoes, tea and tobacco, it may be as low as 0·5, whereas for brands including sales at individual shops, it is certain to be above unity. The reason for this is simple: brands and shops compete with each other, and since the price-elasticity of demand is a measure of competition, it is bound to be higher than it is for a product as a whole, which never has a product so closely interchangeable with it as one brand is with another brand of the same product.

All this may appear to be rather academically theoretical. However, to the pricing policy maker an understanding of the theory of optimum pricing is essential if the aim of his decisions is the maximisation of profits. It is sometimes said that the price-elasticity of demand for an individual brand of a product is unascertainable. With the development of market research in the field of test marketing in recent years and the growing use of mathematics and statistics in industry, it is questionable whether any important item of market information of this kind is unascertainable. In any event, there is not much point in marketing, market research, and test-marketing in particular, if a fundamental piece of information like the price-elasticity of demand cannot be determined.

Undoubtedly, pricing policy decisions are sometimes perforce based on short-term considerations. But whatever the short-term considerations may be, the effect on the company's finances of any deviation of the actual price being charged for the time being from the optimum price should be recognised for what it is—a reduction of net profit. Short-term considerations may dictate a price below the optimum. It is difficult to imagine any circumstances in which they could dictate a price above the optimum.

12

Index numbers[1]

Enough has been said in the foregoing chapters to indicate that index numbers are a subject that no management statistician can afford to neglect. Most emphasis has been placed on the practical value of labour-weighted production index numbers; but price index numbers also have many an important role to play in business management. A good price index of materials used in a manufacturing establishment, or by a building contractor, for instance, can save many man-hours of detailed costing. A good value-weighted quantity index of materials purchased or consumed can provide a useful check on purchases and consumption.

It often happens that the simplest formula turns out to be the most logical and workable. This, at any rate, is true of index numbers. There was a time when some official index numbers were the weighted geometric means of prices or quantities. The index of wholesale prices is an example. Soon after the Second World War, official index numbers were recalculated on the basis of a very simple arithmetic mean formula. If one bought a basket of goods on, say, 31 December 1960 for £5 and a precisely similar basket on 31 December 1962 for £5.50, the price index for those goods at the end of 1962 compared with the end of 1960 would be 110, that is:

$$\frac{5{\cdot}50}{5{\cdot}00} \times 100$$

Fundamentally, the modern official formula is as simple as that.

Some hundreds of index formula have been advocated from time to time over the years and all fail in one respect or another, even the official one, which, under the influence of the Statistical Office of the

[1] The early sections of this chapter consists in part of my article on 'Index Numbers in Business', *The Accountant*, 28 August 1965. By courtesy of the Editor.

United Nations Organisation, has now been adopted almost universally. Some theoretical statisticians object to index numbers on the ground that they cannot be made to conform to the theory of error, and one or two have in consequence declared they will not even discuss them. One even described them as 'this academic tomfoolery'. This, of course, is indefensible. Index numbers have too great a practical value for them to be dismissed in such a fashion.

Geometric formulae, of which there have been some enthusiastic advocates, mostly fail because a relatively unimportant item can have an important effect on the result. This can be demonstrated with a very simple example. If a basket of goods contains three items, one costing 20, the second 8 and the third 2, total 30, in the base period, and 20, 8 and 1, total 29, in the given period; then the arithmetic mean index in the given period (base period = 100) is:

$$100\frac{29}{30}=96{\cdot}7$$

The geometric mean index for this example is based on the cube root of the product of the three costs in each period, that is:

$$100\frac{5{\cdot}43}{6{\cdot}84}=79{\cdot}4$$

This seems much too low; but the absurdity of the geometric mean is seen when a price or quantity falls to zero. Then since the product of any series of numbers of which one is zero, is itself zero, the geometric mean index too is zero, no matter how the other prices or quantities behave.

In practice, it is usually convenient for various reasons to regard the values or costs in the base period as weights. Suppose, in the example we have just used, the quantity, and the prices in the two periods, are as follows:

		Base period		*Given period*	
Item	*Quantity* (1)	*Price p* (2)	*Value p (1)×(2)* (3)	*Price p* (4)	*Value p (1)×(4)* (5)
A	5	4	20	4	20
B	4	2	8	2	8
C	1	2	2	1	1
	Total value		30		29

The unit of quantity can be different for different items, but once a unit has been adopted for an item it should not be changed. Using the values in the base year, we should need to apply them to what is known as the price relatives. The price relative is the ratio of the price in the given period to the price in the base period. For the example we have:

Item	Price relative
A	$\frac{4}{4}=1$
B	$\frac{2}{2}=1$
C	$\frac{1}{2}=0{\cdot}5$

These are now multiplied by the corresponding value in the base period, and the total of these divided by the total value in the base period, that is:

$$\frac{20\times 1+8\times 1+2\times 0{\cdot}5}{20+8+2}=\frac{29}{30}=0{\cdot}967$$

to base = 1·00, or 96·7 to base = 100. For the rest of this chapter, it will be convenient to adopt the convention generally used in discussing index numbers of taking the base as 1·00, as distinct from the convention used in presenting index numbers of taking the base as 100.

LASPEYRES'S INDEX NUMBER

That the formula used with the price relatives is the same as the direct value ratio formula can be proved mathematically. Let Σ indicate summation; subscripts n and o represent the given and base periods respectively; p, the price, and q the quantity, so that p_n and q_n are the price and quantity in the given period; and p_o and q_o, the price and quantity in the base period, then if we use $P_{n.o}$ to represent the price index in the given period relative to the base period (and $Q_{n.o}$ the quantity index in the given period relative to the base period), we have:

$$P_{n.o}=\frac{\Sigma\left(q_o p_o \frac{p_n}{p_o}\right)}{\Sigma q_o p_o} \qquad \ldots \text{(i)}$$

$$=\frac{\Sigma q_o p_n}{\Sigma q_o p_o} \qquad \ldots \text{(ii)}$$

Equation (i) is frequently called an arithmetic formula, and equation (ii) an aggregative formula, although in fact, in this case, they are identical. Similarly, for the quantity index with value weights:

$$Q_{n.o} = \frac{\Sigma\left(p_o q_o \dfrac{q_n}{q_o}\right)}{\Sigma p_o q_o} \qquad \ldots \text{(iii)}$$

$$= \frac{\Sigma p_o q_n}{\Sigma p_o q_o} \qquad \ldots \text{(iv)}$$

The formulae of equations (ii) and (iv) are known as Laspeyres's index numbers, but since equations (i) and (iii) are identical to them the name has come to be generally applied to them also.

Prices vary over time, and so also do quantities purchased or consumed. The consumption of consumer goods varies from time to time owing to changes in taste and relative changes in prices. If, for instance, the price of butter falls and the price of margarine rises, the housewife will tend to buy more butter and less margarine. An increase in real incomes would have a similar effect: more butter and less margarine; more cake and less bread; and more bread and butter with jam on it. The formulae of equations (iii) and (iv) are therefore not simply of academic interest.

PAASCHE'S INDEX NUMBER

It may be asked why should we not use given period value weights instead of base period weights? If quantities vary, would not the answer for a price index be different? Generally, but not quite always, the answer would be different; and this form of index, which is called Paasche's index number, is just as good as Laspeyres's, but it is less convenient in practice. The weights used in Laspeyres's index are the same in the successive periods, whereas in Paasche's index, they change in each period. Furthermore, the data are not always readily available. Weighting data for some official index numbers are available only for years in which a detailed Census of Production is taken, such as 1954, 1958, 1963 and 1968, and that is why these index numbers are based on one or other of these years.

It could be argued that if Laspeyres and Paasche give different results, one or both must be biased. This is perfectly true: in fact, both are biased, one upward and the other downward. Where the quantities have changed owing to relative price changes, so that prices have increased by more than average and quantities have decreased by more than average, Laspeyres has the upward bias and Paasche the downward bias in both price and quantity index numbers. But it is worth noting here, that where the quantities or the prices or both remain constant or change in the same direction and in the same ratio, Laspeyres and Paasche give the same, and no doubt the correct, answer for both the quantity and price index.

Apart from the matter of practical convenience, Paasche has the same shortcomings as Laspeyres. Paasche's aggregative formula for the price index is:

$$P_{n.o}=\frac{\Sigma q_n p_n}{\Sigma q_n p_o} \qquad \ldots (v)$$

and the corresponding arithmetic formula known more correctly as the harmonic, is:

$$P_{n.o}=\frac{\Sigma q_n p_n}{\Sigma\left(q_n p_n \dfrac{p_o}{p_n}\right)} \qquad \ldots (vi)$$

It will be observed that the formula of equation (vi) takes values in the given period as weights. For this reason it is of greater practical value than equation (v), which assumes we know the quantities in the given period for revaluation at base period prices.

The same can be done for Paasche's quantity index:

$$Q_{n.o}=\frac{\Sigma q_n p_n}{\Sigma q_o p_n}$$

To avoid the danger of assuming that we have quantity data in each period, we use the price relatives in the harmonic version:

$$Q_{n.o}=\frac{\Sigma q_n p_n}{\Sigma\left(q_o p_o \dfrac{p_n}{p_o}\right)} \qquad \ldots (vii)$$

This looks more like a price index than a quantity index, but it is a quantity index for all that. For multiplied by Laspeyres's price index, equation (i), it gives the value index, which is a subject of the next paragraph.[1]

[1] It is worth mentioning that the assumption of the availability of quantity data in each period can similarly be avoided in Laspeyres's quantity index, equation (iii):

$$Q_{n.o}=\frac{\Sigma\left(\dfrac{p_o}{p_n} p_n q_n\right)}{\Sigma p_o q_o}$$

Laspeyres and Paasche are so closely related that the several formulae are known as the formulae of the Laspeyres–Paasche group. This can best be demonstrated by relating the index number of prices or quantities to the value index, $V_{n.o}$, where

$$V_{n.o} = \frac{\sum q_n p_n}{\sum q_o p_o}$$

Now if we divide $V_{n.o}$ by Laspeyres's price index, we would expect to arrive at a quantity index and so we do:

$$Q_{n.o} = \frac{\sum q_n p_n}{\sum q_o p_o} \cdot \frac{\sum q_o p_o}{\sum q_o p_n}$$

$$= \frac{\sum q_n p_n}{\sum q_o p_n}$$

but, oddly enough, it is Paasche's quantity index; from which it follows that Laspeyres's price index (quantity index) multiplied by Paasche's quantity index (price index) always gives the value index.

FISHER'S IDEAL INDEX NUMBER

As a result of this relationship, and the apparently opposite direction the bias takes, Irving Fisher, at one time the leading authority on index numbers, advocated the geometric mean of the two formulae as the ideal. This geometric mean came to be known as Fisher's Ideal.[1] For a price index number, it is:

$$P_{n.o} = \sqrt{\frac{\sum p_n q_o}{\sum p_o q_o} \cdot \frac{\sum p_n q_n}{\sum p_o q_n}} \qquad \ldots \text{(viii)}$$

with the aggregative formulae. Laspeyres's arithmetic and Paasche's harmonic can, of course, be used in place of the aggregative formulae.

Fisher's Ideal price index divided into the value index gives Fisher's Ideal quantity index, so that it can be said that for Fisher's Ideal:

$$P_{n.o}\, Q_{n.o} = V_{n.o} \qquad \ldots \text{(ix)}$$

This is known as the factor-reversal test, which any perfect index formula should satisfy. Neither Laspeyres nor Paasche satisfy it.

A similar kind of test is the time-reversal test, which requires that:

$$P_{n.o}\, P_{o.n} = 1{\cdot}0$$

[1] Fisher, *The Making of Index Numbers* (1922).

Again, Fisher's Ideal satisfies this test, but not Laspeyres or Paasche. For Laspeyres's price index, for instance:

$$P_{o.n}=\frac{\sum p_o q_n}{\sum p_n q_n}$$

which is the inverse of Paasche. Since Laspeyres and Paasche do not usually give the same answer, one divided by the other cannot usually equal unity. For Fisher's Ideal, we have:

$$(P_{n.o}\,P_{o.n})^2=\frac{\sum p_n q_o}{\sum p_o q_o}\cdot\frac{\sum p_n q_n}{\sum p_o q_n}\cdot\frac{\sum p_o q_n}{\sum p_n q_n}\cdot\frac{\sum p_o q_o}{\sum p_n q_o}$$

which all cancel out to give 1·0. The root of 1·0 is 1·0, so that Fisher's Ideal satisfies the time-reversal test. It will be observed that to arrive at $P_{o.n}$, subscript n is substituted for o, and o for n throughout.

On the score of convenience and practicability, however, Laspeyres's formula is still the best of the three. Fisher's Ideal not only requires twice the amount of arithmetic, which in these days of electric calculators and electronic computers may not matter very much; but it also requires all the data necessary for Paasche, and that is something of a drawback as all practical makers of index numbers know.

In order to keep the amount of work within reasonable limits, many index numbers are based on a sample of the population of which the index is designed to provide a measure. The sample may be roughly stratified in the sense that it contains all the major items, and only a proportion of the remainder. With a price index, grossing-up is unnecessary. But with a quantity index, grossing-up may be essential in some cases.

The kind of data available regularly, say monthly or quarterly or even annually, is often limited to prices of most items, and the total values of all items. Owing to the lack of quantity data, the usual procedure for obtaining a quantity index is to divide the total value index by the price index. As we have seen, if the price index is Laspeyres's, the resulting quantity index is Paasche's. But since Paasche is as good as Laspeyres, with the bias in the opposite direction, it is of no consequence. The process of dividing the value index by the price index is often called *deflating*, which in these days of rising prices is good enough: the quantity index always tends to stand lower than the value index. But if the prices were falling, the process would have to be called *inflating*, since in that event the quantity index would tend to stand higher than the value index.

Whether the price index is calculated from a sample or from the total, the quantity index derived by dividing the value index by the price index does not need grossing up; and it is as representative as the price index no more and no less.

REPRESENTATIVENESS

To what extent a price index based on a sample is representative of the total depends upon the dispersion of the price-relatives in the *n*th period. Suppose our sample consists of five items, then to obtain a measure of the dispersion, we proceed as shown in Table 12.1.

TABLE 12.1. Testing for representativeness of sample

	Price		*Price relatives*		
Item	*In base period* p_o	*In given period* p_n	$\frac{p_n}{p_o}$	*Deviation from mean*	*Deviation squared*
A	48	48	1·0	0·0	0·0
B	30	36	1·2	0·2	0·04
C	85	51	0·6	−0·4	0·16
D	100	140	1·4	0·4	0·16
E	46	37	0·8	−0·2	0·04
TOTAL			5·0	1·2[1]	0·40
Average per item			1·0	0·24	0·08

Standard deviation $= \sqrt{0 \cdot 08} = 0 \cdot 283$

[1] Ignoring sign

In this example, the average price relative is 1·0, the average deviation is 0·24, and the standard deviation 0·283, both of which are measures of the dispersion. The dispersion, it can be said, is too great for the index number to be regarded as reasonably accurate. The sample of five items is much too small, and if more items are available, the sample should be increased to make the index number more nearly representative.

SUB-INDEX NUMBERS

Most index numbers can be made up of sub-index numbers. Indeed, in practice, the usual procedure is to calculate the appropriate sub-index numbers first, and then to amalgamate them to give the overall index number. The advantage of this is that each sub-index can be given its full weight in the amalgamating process independently of the size of the sample used in arriving at the sub-index. For some sections of a population or field, a representative sample may be obtained from a smaller proportion of the section total than with other sections. In order to give each section its proportionate weight in the overall index, the approach to the overall index *via* section sub-index numbers is essential. Where all items in the population are used, sub-index numbers are not necessary for calculating the overall index; but sub-index numbers are often useful in themselves for various purposes, such, for instance, as discovering which sections of the stores show the greatest increase in costs which is particularly important where substitution is possible.

SPLICING ON AND OFF

Changes in quality present a problem to index number makers. If, for instance, a car manufacturer introduced an exhaust system made of a moisture-resisting metal such as copper or stainless steel in place of the ordinary mild-steel type, which rusts quickly, the index number maker concerned with a price index of stores would have to decide whether to leave the original value and price of exhaust systems in the base period or increase them to a level that would take account of the higher cost of the new type. Both the weight and the price relative would be involved if he decided to adopt the higher levels. But would it be right? The answer must depend on the purpose of the index. Generally, it would be; but the users of the index should be informed of the change.

The price and value of the new item in the base period is often unknown in changes of this kind. When this is the case, the only thing that can be done is to estimate them on the basis of the trend of the old item. The process is quite simple. If the price relative of the old item has risen to 1·2 in the given period compared with the base, then the new item will be introduced into the index at this price-relative in the given period so that if the price of the new item in the given period is £12, its price in the base period would be assumed to be £10, and the value of the item in the base period for weighing purposes would be raised to correspond to this price. This would preserve the comparability and continuity of the index.

What we have done in effect is to drop the old item and introduce the new one. The process of introducing a new item whether in substitution or in addition, is called *splicing on.* Dropping an old item that is no longer contained in the population is often a matter of excluding it from the calculations for the given period, which should automatically exclude its weight in the base period. Where this happens, one point to watch is that the representativeness of the index or sub-index is not adversely affected by the loss of the item.

Splicing on is a technique which most index-number makers have to learn at some time in their careers. For a substitute item the price-relative of the displaced item in the given period can be used for estimating the appropriate base period price and value, as described above. But for an additional item which it is now thought important enough to include in the sample, and for which there is no actual base period price, the sub-index number of the section to which it belongs has to be used for estimating the base period price of the new item. This has the effect of introducing the new item in the given period at a price relative equal to the sub-index of the section, which seems reasonable enough. Even where an actual base period price of the new item is available, it could still be the best method, since if there is a base period price, it is safe to suppose that there are prices available for all intervening periods between the base and the given periods. If one used the actual base period price, it would be necessary in order to preserve comparability to introduce these prices and to amend the sub-index and total index for all intervening periods, which since the item is a relatively new addition to the population would not only be superfluous but also illogical.

Preserving comparability is all-important. Strictly, an index number of the kind so far discussed, can be compared with its base only. To compare the index numbers for any two periods neither of which is the base is not strictly correct. However, it is usually quite harmless, and is frequently done. If one could not compare $P_{n.o}$ with $P_{(n-1).o}$ or $P_{(n-2).o}$ the making of index numbers would not be worth while.

THE CIRCULAR TEST

An index number in which the index for any period can strictly be compared with the index for any other is one that will pass what is called the circular test. The circular test is in theory an important one, yet oddly enough, unless the accepted ideas of a properly weighted index formula are abandoned, the test passes many bad formulae, and condemns all good formulae, with one exception, as bad. The exception is a cross-weight form of Laspeyres:

$$P_{n.o}=\frac{\Sigma q_a p_n}{\Sigma q_a p_a} \qquad \ldots \text{(x)}$$

in which a denotes the arithmetic means of the quantities and prices of each item in all periods. It is probably the best index for use in a closed system for the purposes of correlation and similar analyses, but it would be of little if any use in a continuous index designed to run for an indefinite number of periods into the future, like most official index numbers. Divided into its corresponding value index,

$$\frac{\Sigma q_n p_n}{\Sigma q_a p_a}$$

it produces its Paasche counterpart of the quantity index, so that in this respect it behaves like the standard Laspeyres. This form of Paasche does not satisfy the circular test, since the value weight becomes $q_a p_n$, and not the sum of the averages. Equation (x), an aggregative, has its arithmetic form:

$$P_{n.o}=\frac{\Sigma q_a p_a \dfrac{p_n}{p_a}}{\Sigma q_a p_a} \qquad \ldots \text{(xi)}$$

Where only values are available for weighting purposes, this is the formula to use in a closed system for regression analyses and similar work requiring strict circular-test comparability.

USES OF CHAIN-BASE FORMULAE

In some fields, items tend to change in design and specification, so frequently that in course of time a representative sample of items with comparable base period and given period data cannot be drawn from the total. This is liable to happen where the items consist of complex manufactures such as machinery, and vehicles, and assemblies and sub-assemblies for durable consumer goods, for which styling and design are important selling points. It is always possible, however, to draw a representative sample with comparable data for two consecutive periods. When this difficulty arises the so-called chain-base form of

Laspeyres (or Paasche for that matter) can be used. Laspeyres's chain-base aggregative formula is:

$$P_{n.o}=P_{(n-1).o}\cdot\frac{\sum q_{n-1}\,p_n}{\sum q_{n-1}\,p_{n-1}} \qquad \ldots \text{(xii)}$$

and the arithmetic:

$$P_{n.o}=P_{(n-1).o}\cdot\frac{\sum q_{n-1}\,p_{n-1}\,\dfrac{p_n}{p_{n-1}}}{\sum(q_{n-1}\,p_{n-1})} \qquad \ldots \text{(xiii)}$$

This formula appears to refer back to the original base period, and it is true that the index in every period is standardised to the base period = 1·00. However, since the weights used in the arithmetic chain index are those of the previous period, strict comparability is limited to two consecutive periods, and not between any given period and the base. Reference to the base = 1·00 is a convenience, and is made, as equation (xiii) shows, by multiplying $P_{n.n-1}$ by $P_{(n-1).o}$, so that the index for the given period is the product of the index numbers each derived from consecutive periods, that is:

$$P_{n.o} = P_{1\cdot 0}.P_{2\cdot 1}.P_{3\cdot 2}\ldots P_{(n-1)\cdot(n-2)}.P_{n.(n-1)} \qquad \ldots \text{(xiv)}$$

If, for instance, the given period is the third, counting the base as zero, we might have something like this:

	Index
Base	= 1·00
$P_{1\cdot 0}$	= 1·20
$P_{2\cdot 1}$	= 1·35
$P_{3\cdot 2}$	= 1·40

Then $P_{3\cdot 0}=1{\cdot}20\times 1{\cdot}35\times 1{\cdot}40=2{\cdot}268$

In practice, $P_{2\cdot 0}(=1{\cdot}62)$ would already have been calculated, so that for the third period we would merely have $1{\cdot}62\times 1{\cdot}40$ to calculate. Hence the introduction into equation (xiii) of the index for period $(n-1)$ to the base = 1·00, i.e. $P_{(n-1).o}$. It could just as easily have been all the terms (except the last) of equation (xiv).

Perhaps a word of warning is necessary. If price changes tend to take place more often with changes in specification than otherwise, the selection of a new sample in each period, which would necessarily exclude items whose specification has changed, would—as it has been described—be throwing out the baby with the bathwater.

In spite of what some academic statisticians think, say, or believe about them, index numbers have become an essential tool of business management and government. Index number makers necessarily have to know something about the theory of index numbers as well as the techniques used in the best practice of making them.

Index number users, too, need to know something of these matters if they are to make the best use of them, especially as there is no such thing as the perfect index number in the sense that there is a perfect and exact answer to two times two. Laspeyres and Paasche and even Fisher's Ideal provide criteria rather than perfection. As explained above, all three give the perfect answer in certain conditions rarely found in practice, and no other known index formulae do this.[1] That the three give different answers—albeit often only slightly different—when these theoretical conditions are not satisfied, is evidence that at least two of them are not perfect. Undoubtedly, Fisher's Ideal is the nearest to perfection.

BIAS

Laspeyres provides a sound criterion, far superior to any other practicable formulae. Relative to Paasche and therefore to Fisher's Ideal, it often, but not always, has an upward bias, but it could never be described as erratic.

The upward bias appears in working backward from the base period as well as forward. Indeed, in working backward from, say, 1960=1·00, with 1960 weights, to 1956, one is in effect using Paasche; that is, rebasing to 1956=1·00 (but still with 1960 weights), there would be a tendency to a downward bias in 1960 compared with 1956. In short, there may be a break in trend comparability at 1960, since at that date there is a change in criterion. A simple example will show what might happen where the bias is considerable.

Index, 1960=1·00	*1956*	*1960*	*1963*
Laspeyres (1960 weights)	1·00	1·00	1·20
Paasche	0·90	1·00	1·10
Fisher's Ideal	0·95	1·00	1·15

[1] There is, perhaps needless to say, a fourth, and that is another cross-formula index consisting of the arithmetic mean of Laspeyres and Paasche.

The weights used for the Paasche index for 1956 would be those of 1956; and for 1963, those of 1963. If we take Fisher's Ideal as the correct index—and it would certainly be free of any built-in bias—we can see at once that there is a risk in using Laspeyres (or Paasche) backward as well as forward.

In this example the bias is exaggerated beyond the possible. There is a limit to the bias that can exist. The bias arises where there is any degree of correlation between the quantity relatives and the price relatives. When there is any correlation at all, it is nearly always negative, and this gives Laspeyres an upward bias compared with Paasche. The limit to bias is set by a set of price and quantity relatives that are perfectly correlated negatively, that is, where the coefficient of correlation equals $-1{\cdot}0$, which can occur only where price is allowed to influence the quantity in conformity to the law of supply and demand. In conditions, such as a factory store, under which relative changes in prices do not affect the relative consumption of the several materials, there is little or no correlation, and therefore little or no bias. Indeed, if the consumption of the several materials remains in the same proportion and there are no specification changes, Laspeyres and Paasche would necessarily give the same answer, and it would be the correct answer.

PRODUCTIVITY INDEX NUMBERS[1]

It is probably true to say that all index numbers are ratios standardised to a base. The base is not necessarily a point of time or a period: it can be a place such as a country or a factory or a company of a group. International organisations sometimes make index numbers of retail prices for a given time relating prices in various countries to prices in one used as base. Similarly, it may be possible to compare productivity in a group of factories or companies with one of them used as base.

However, the difficulties of making index numbers of this kind are considerable. They are typified by those of making an international price index. Here, the average housewife's basket of goods varies from one country to another, so much so that the only feasible formula to use is the crossweight form of Laspeyres of equation (x) in a closed system; and even then there remains the difficulty of products and

[1] This section is based in part on my article 'Measuring Productivity: the Delusion of Value Added', *The Certified Accountants' Journal*, February 1971. By courtesy of the Editor.

services common in some places and unknown in others. But there is no need to go outside England to meet this problem. There are food products which are consumed in large quantities in the Midlands and North of England but which are almost unknown in the South: cow heel, black pudding, called pig pudding in some localities, hodge and chicklen, tripe, savoury ducks, known as faggots in the West Midlands, and oatcakes (English version). Anyone attempting to compile a cost of living index comparing the various counties in England would not find it so easy as making an index comparing different periods for one county.

It is suggested that any attempt to make a productivity index comparing different factories or companies at the same time would meet with failure for much the same reasons. This is not to say that one could not make a productivity index for each factory and compare their several trends one with another. A comparison of the several index numbers for a given period would not indicate the relative efficiencies in the factories; it would merely show the relative progress made in them since the base period.

In this section, we are concerned with making productivity index numbers for different periods and for the same place. As with price and quantity index numbers, they are based on ratios, which in turn call for the determination of two figures: a numerator and a denominator. With Laspeyres, the denominator is the value in the base period and it remains constant, the numerator being the variable. With Paasche both are variables. We have seen that a quantity index derived by dividing the value index by Laspeyres's price index is Paasche's quantity index. This applies to the making of productivity index numbers, with a labour-weighted production index as the numerator and the ratio of labour consumed or used in the given period to that in the base period as the denominator. First, we consider the numerator.

The labour-weighted production index

It has long been thought that the overriding problem of making productivity index numbers is the need to eliminate the effect of changes in make or buy. It was believed that the use of value added would provide the necessary solution, but it has been found that it goes only part way to a solution. It appears that value added as a numerator eliminates only a proportion of the effect of changes in make or buy depending upon the proportion of value added consisting of wages and the proportion consisting of profits. The logic of the argument is: when a change in make or buy takes place, then given there are no significant differentials in wage rates, the wages bill itself rises or falls *pro rata* to the labour employed. Since the wages bill forms only a part of the value added, value added is bound to rise or fall proportionally less than the labour force. Other things being equal, a change from make to buy would therefore cause a rise in an index of productivity based on value added, and a change from buy to make would cause a fall.

It follows that a productivity index must have as its numerator a production index built up in detail with additional items such as result from a change in buy to make spliced on, and discarded items (make to buy) spliced off. A production index has the additional attraction over value added of being applicable to detailed processes and products, which renders the resulting productivity index of greater use in productivity bargaining and for other purposes calling for detailed figures.

A simple hypothetical example can be used for demonstrating the statistical calculations. Suppose a jam factory has three main processes:

A. Making, including preparing fruit and boiling;
B. Filling into pots each containing 1 lb of jam; and
C. Capping the pots, labelling, and packaging ready for distribution.

In 1969, the factory makes, fills, caps, etc., 100,000 lb of jam. In 1970 advanced filling and capping technologies are introduced with the result that there are considerable savings in manpower in 1971. There is no difficulty in ascertaining the productivity at the process level. If the factory makes 150,000 lb in 1971, and the labour force increases from 40 men in 1969 to 60 men in 1971, the productivity in the making department remains the same. If the labour force averages 50 men in 1971, then there is an increase in productivity of 20 per cent, the index number for 1971 being 120 to the base 1969 = 100:

$$\frac{150}{100} \div \frac{50}{40} = 1{\cdot}20$$

Suppose the firm has overestimated the demand in 1971, and although 150,000 lb of jam are made and filled into pots, the factory caps, labels, etc. and sells only 120,000 lb in that year. Table 12.2 gives the hypothetical detail and the calculations needed to provide an overall labour-weighted production index for the factory. The ratios given in the last

TABLE 12.2 A labour-weighted production index

Process	1969			1971				
	Output lb '000	*Man-years No.*	*Man-years per thou lb*	*Output lb '000*	*Man-year No*	*Man-years per thou lb*		
	q_o	m_oq_o	m_o	q_n	m_nq_n	m_n	m_oq_n	$\frac{m_o}{m_n}$
A	100	40	0·400	150	50·0	0·3333	60·0	1·20
B	100	10	0·100	150	6·0	0·0400	15·0	2·50
C	100	3	0·030	120	1·5	0·0125	3·6	2·40
TOTAL (Σ)		53			57·5		78·6	

Production index $Q_{n \cdot o} = \frac{\Sigma m_o q_n}{\Sigma m_o q_o} = \frac{78 \cdot 6}{53 \cdot 0} = 1 \cdot 483$

column of Table 12.2 provide the productivity index numbers process by process.

The labour ratio as denominator

For the three processes detailed in Table 12.2, the appropriate labour ratio is $\sum m_n q_n / \sum m_o q_o$, which corresponds to the value ratio, where p serves for m. For the example, the ratio is 57·5/53·0=1·085, which is divided into the production index to give the productivity index. However, if the productivity index for the factory as a whole including warehousing labour and office staff is required, the total labour forces, including these people, in the given and base periods would be used for arriving at the labour ratio to use as the denominator.

The productivity index

Let the productivity index in the given year to 1·00 in the base year be represented by $M_{n.o}$, then, for the three departments or processes, we have:

$$M_{n.o} = \frac{1{\cdot}483}{1{\cdot}085} = 1{\cdot}367$$

However, since the ratio m_o/m_n gives the productivity index process by process, one might expect that weighted for output they would give a productivity index for the three processes combined. And so they do. The index derived from the weighted figures is:

$$M_{n.o} = \frac{\sum m_o q_n}{\sum m_n q_n} = \frac{78{\cdot}6}{57{\cdot}5} = 1{\cdot}367$$

which is exactly the same as the figure above derived by dividing the labour ratio into the production index.

We have the labour weighted production index (see Table 12.2):

$$Q_{n.o} = \frac{\sum m_o q_n}{\sum m_o q_o} \qquad \ldots \text{(xv)}$$

and the labour ratio ($L_{n.o}$):

$$L_{n.o} = \frac{\sum m_n q_n}{\sum m_o q_o} \qquad \ldots \text{(xvi)}$$

Equation (xv) divided by equation (xvi) gives:

$$M_{n.o} = Q_{n.o}/L_{n.o} = \frac{\sum m_o q_n}{\sum m_o q_o} \cdot \frac{\sum m_o q_o}{\sum m_n q_n} \qquad \ldots \text{(xvii)}$$

The term $\sum m_o q_o$ cancels out leaving:

$$M_{n.o} = \frac{\sum m_o q_n}{\sum m_n q_n} \qquad \ldots \text{(xviii)}$$

which is the same as the above. It will be observed that it is, in effect, the inverse of Paasche's price index weighted for quantity with m serving for p.

However, equation (xviii) is limited in its application to the derivation of productivity index numbers in respect of the labour force accounted for in the process calculations. If, as suggested above, an overall productivity index for the factory is required, then it would be necessary to apply equation (xvii) in which the labour ratio is the total factory labour force in the base period divided by that in the given period. If for our hypothetical jam factory, the two labour figures are 90 and 100 respectively, the overall productivity index would be:

$$90/100 \text{ of } 1{\cdot}483 = 1{\cdot}335$$

The example introduces another problem, and solves it, that of a case in which a product is stored awaiting final processing and delivery in the following year. Value added could not take care of the labour expended on the surplus going into store, even if it were otherwise satisfactory in providing a valid numerator for an overall index of productivity.

There is a common misconception about productivity index numbers. It is thought that a rise in productivity resulting from the introduction of an advanced technology should be ascribed to the new technology by using it as the denominator instead of labour. How this can be done is not clear. What statistical figures can be used to provide a technology ratio? The argument is that by using a labour ratio as the denominator, the index number maker is attributing the rise in productivity to labour. This is not so. A productivity index number is a statistical entity like a business ratio. A productivity index number no more attributes a rise in productivity to labour than the ratio of turnover to share capital attributes a rise in that ratio to shareholders. Many ideas and proposed methods have been adduced to solve the problem. But the fact of the matter is there is no problem. Most new technologies are designed to save labour. What better measure of their success could there be than a productivity index on the lines demonstrated above, short of making a complete financial assessment?

It is not possible here to range over the whole subject of index numbers; it is hoped that the foregoing deals with all the more practical problems and pitfalls that are likely to confront index-number makers in the management field, and to give index-number users a clear understanding of the nature of index numbers. Many books have been written on the subject, some extolling the virtues of this or that formula. Below is given a short bibliography for the benefit of anyone who wishes to go more fully into this fascinating subject.

1. Irving Fisher, *The Making of Index Numbers*, Boston, Mass, 1922.
2. R. Stone, *Quantity and Price Indexes in National Accounts*, Paris, OEEC 1956.
3. Ministry of Employment, *Methods of Construction and Calculation of the Index of Retail Prices*, HMSO 1956.
4. Central Statistical Office, *The Index of Industrial Production, Method of Compilation*, HMSO 1959.
5. Croxton and Cowden, *Applied General Statistics* (Chapter on Index Numbers), Pitman 1955.

No. 3 demonstrates the splicing on and off processes for seasonal items such as strawberries; and No. 4 discusses and demonstrates the use of indicators where valid production statistics are not feasible (the consumption of newsprint to represent the production of newspapers is an example). It also indicates how resourceful index number makers have to be.

13

Trend forecasting

Few if any of us have the gift of prophecy. It is safe to say, indeed, that prophecy, divination or forecasting have in the past been so closely associated with superstition, that to profess to have the gift nowadays is to invite a smile of pity. In a footnote under reference 511, *Prediction*, Roget in his *Thesaurus* gives a list of more than 50 terms 'expressive of different forms of divination'. The temptation to quote a few of them cannot be resisted:

Theomancy, *by oracles*;
Cristallomantia, *by spirits seen in a magic lens*;
Aeromancy, *by appearances in the air*;
Genethliacs, *by the stars at birth*;
Meteoromancy, *by meteors*;
Alectryomancy, *by a cock picking up grains*;
Pegomancy, *by fountains*;
Geomancy, *by dots made at random on paper*;
Catoptromancy, *by mirrors*;
Palmistry, Chiromancy, *by the hand*;
Arithmancy, *by numbers*;
Ceromancy, *by dropping melted wax into water.*

It will be inferred that prophecy is an ancient profession. However, its foundation is not entirely one of superstition. Much forecasting such as of the times of sunrise and sunset, eclipses of the sun and moon, the tides and the Severn bore have a foundation in facts observed over long periods. Nobody can hope to achieve such a high measure of accuracy in the business field of demand, cost and revenue forecasting. But this is not to say that we must have recourse to *theomancy*, *cristallomantia* or any other fancifully superstitious method. We may use numbers, and dots made on paper, but the numbers will be relevant, not omens, and the dots made on paper will be scientifically located and not made at random.

In the sphere of business management there are no absolutes—no absolute truth, no absolute right, no absolute wrong. This goes for the field of forecasting more than for any other. If there is an absolute truth, it is that there is no other. One sphere within the scope of human understanding where absolute truths and untruths exist is that of pure mathematics. The application of mathematics to the practical problems of business management has the immediate effect of introducing an element of uncertainty.

The use of mathematics applied to reasonably good basic data has reduced the hazard in business forecasting. Nevertheless, it remains an extremely hazardous process. Important as method may be, the data are more important. The most mathematically sophisticated method in the world is no substitute for facts. The actual data available relate to the present and past, and never to the future. It is true that there are expert estimates of future data such as the Treasury's estimates of the GNP, from which estimates of practical value to management statisticians can be derived, items such as the total personal income. But the question is whether expert estimates are reliable. Consider the GNP for instance. The actual figures for the years 1959–69 at current prices as published in the *National Income* blue book for 1970, are given in Table 13.1 together with the year-by-year absolute and percentage increases, and the corresponding percentage increases at 1963 prices.

TABLE 13.1 The gross national product at factor cost

	Total for year	*Year-by-year increase*		
		(at current prices)		*(at 1963 prices)*
	£'mn	*£'mn*	%	%
	(1)	(2)	(3)	(4)
1959	21,422	—	—	—
1960	22,817	1,395	6·51	4·71
1961	24,414	1,597	7·00	3·59
1962	25,556	1,142	4·68	1·30
1963	27,150	1,594	6·24	4·00
1964	29,319	2,169	7·99	5·22
1965	31,342	2,023	6·90	2·78
1966	32,991	1,649	5·26	1·76
1967	34,817	1,826	5·53	1·78
1968	36,819	2,002	5·75	2·84
1969	38,601	1,782	4·84	1·34

One would imagine that the Treasury could make reasonably accurate estimates of the increase in GNP at 1963 prices from these data, and the data it has internally. But neither the Treasury nor any other body of experts has so far succeeded in regularly keeping its forecasts of increases within 25 per cent of what turns out to be the true figure. However, as statisticians, we ought to appreciate the way a percentage like this exaggerates the error. Consider, for instance, the increase at current prices of £1,782 million in 1969. An error of £500 million or about 28 per cent of the increase, would represent an error of about 1·3 per cent of the total for the year. If the forecasts of the values of all relevant factors in demand contained an error of no more than, say, 2 per cent, sales forecasting would be a much less hazardous process than it is.

Trend forecasting ignores the hundred and one short-term factors that are liable to disturb any forecast. Most short-term factors cannot be foreseen, and where they can, an estimate of their effect on a trend forecast is scarcely possible. There are some longer term factors that cannot be foreseen. Changes in Government fiscal and economic policies are cases in point: they are as unpredictable as the vernal flight of the plover. There was a time in the nineteen-fifties when Remploy, a Government sponsored undertaking, whose purpose is to provide employment for disabled people, considered the advisability of closing its furniture factories because of the uncertain incidence on its revenues of changes in purchase tax.

ANNUAL COSTS

Forecasting costs is largely a matter of synthesis. A model is necessary but a mathematical model could scarcely serve any useful purpose. The mathematical cost models given in the preceding chapters are not forecasting equations, they are designed for cost analysis. A numerical model is all that is necessary, its object being to prevent any cost from being overlooked, and to preserve continuity.

Before forecast total annual costs can be compiled, the cost analyst needs to know the forecast sales. And before the statistician can forecast sales he must have the forecast marginal cost of the brand under review. The first step to be taken then in forecasting annual costs and sales, is to make forecasts of the marginal cost for the years to be covered by the forecast. It would scarcely be possible to proceed by denotation, so that the cost analyst would have to use a connoted definition of marginal cost and adjust his results for the residual factors. Direct labour, materials, fuel and power are the items which can usually be taken care of in forecasts of the marginal cost. The analyst should take account of anticipated labour-saving improvements in production and packaging technologies as well as changes in wage rates and materials prices. Index numbers of wage rates and materials prices would be of great value in this respect. It would render less hazardous the simple application of percentage changes to the latest actual marginal cost derived on a denoted definition by analysis.

For the rest, once the statistician has the forecast of sales brand by brand in terms of quantity, he can proceed to build up his forecast of annual costs without much difficulty on the basis of his numerical cost model. For the purpose, the model could usefully embody one or other of the mathematical cost models of equations (xiv) or (xv) of Chapter 5, depending upon whether or not there are any attributable time costs. Equation (xiv) applies where there are no attributable time costs, and the statistician's main effort will be devoted to forecasting the value of the constant F. Where there are attributable time costs that can be treated in the systematic way implied, equation (xv) applies, when the statistician's main effort will be devoted to forecasting the value of F_h. In this equation, b_1, b_2, . . . are constants, and should remain as constants throughout the years of the forecast, so that once their values have been determined, the evaluation of the bN's would be a simple matter. The number of units of attributable capacity, N, is a variable, and its forecast value for each brand would depend in part upon the forecast value of sales and in part upon the capacity currently available.

The numerical part of the annual cost model would be concerned with F or F_h, and would include the following:

- Fees and salaries:
  - Top management:
    - Directors
    - Other
  - Other executives
  - Clerical staff
  - Technical staff
- Indirect wages:
  - Foremen
  - Skilled labour
  - Semi-skilled labour
  - Labourers

Indirect materials
Office expenses:
  Stationery, books and periodicals
  Post and telephones
  Furniture renewals and repairs
Accommodation:
  Rent
  Rates
  Repairs and maintenance
  Cleaning
  Heating and lighting
Machinery, plant and vehicles:
  Annual capital charges
  Repairs and maintenance
  Fuel and power
  Hire charges
Transport and distribution
Miscellaneous:
  Travelling, including hotel expenses
  S.E.T.
  National Insurance
  Payments to pension fund
  Fire insurance
  Redundancy payments

As it stands, this is no more than a skeleton classification of non-attributable annual costs. The flesh consists of details, such as numbers employed or expected to be employed by grade, showing annual salary or wage including annual overtime, bonuses, etc. and the total anticipated payment grade by grade extended in a column on the right.

SALES

There is no alternative to building up total sales brand by brand. The appropriate forecasting model is equation (xvi) of Chapter 5, which relates to an individual brand. When the elasticities of demand have been calculated, then since they can all be assumed to remain constant, the accuracy of forecast sales must depend upon the accuracy of the forecasts of the variables. If the company is pursuing a rational pricing policy and does not permit short-term considerations to interfere with that policy, the accuracy of the forecast of the variable p, the price, will depend upon the accuracy of the forecast marginal cost.

Advertising

The only other variable factor within the full control of the company is its own advertising expenditure. Here, the forecasting statistician should consult the advertising manager, who should be able to say what amount of the company's advertising expenditure applies to each brand. Some advertising may relate to more than one brand; nevertheless, it may be reasonable in some cases to count the whole of such expenditure as chargeable for the purpose to each of the several brands covered. A decision will depend to a great extent upon the scope of the basic data of advertising expenditure used in the regression analysis of the market demand for each brand. If the basic data include the cost of multi-brand advertisements in full, then so should the forecast data.

National income

The rest of the variable factors in demand are all external or environmental. They vary in importance; but consumer income must count as amongst the more important. The statistican can take his choice of the forecasts made by the experts, which include the Treasury, the National Institute for Economic and Social Research, and some of the more serious newspapers and periodicals. Or he can strike an average of them. Or he may decide to make his own forecasts, on the grounds that the experts' forecasts are biased for one reason or another.

Prices of competitors' brands

Past and present competitors' prices may be well known to the company. Nevertheless, any attempt to forecast them is likely to be extremely difficult. Probably the best approach is to try to determine the kind of thinking that lies behind the main competitors' pricing policies at the present time. Do they pursue rational policies ? Do they allow short-term considerations to govern their policies ? Are they seeking by price adjustment a greater share of the market ?

The meetings of the trade association provide one source of information of this kind. A careless word here, a hint there, can give a useful insight into a competitor's thinking. Sometimes there may be a general discussion on market prices in the trade, even now, long after the Monopolies Commission put paid to price rings. Market leaders all hope tacitly that they will retain the leadership in pricing as in other matters and may give more than an inkling of their future pricing policy.

An index of competitors' prices, which would represent an independent variable in the regression analysis of market demand, may throw some useful light on competitors' pricing policies. But however valid the inferences that may be drawn from past behaviour, no valid inference can be drawn in respect of future behaviour. In short, there must always be an element of hunch in forecasting competitors' prices.

Competitors' advertising

Here, there may be scope for simple trend forecasting, i.e. the extrapolation of present trends. The company's advertising manager may be able to help; after all, he has the advertising mind, and is best able to read the minds of other advertising managers. *Ad hoc* advertising and promotion campaigns are necessarily excluded as an independent variable from the regression analysis of the demand for a brand of product or service. They must also be excluded from any trend forecast of competitors' advertising. They count as a short-term disturbing factor. Here again, a good advertising manager who keeps his ear to the ground may be willing to hazard what can rarely be little better than a guess about future advertising and promotion campaigns to be conducted by competitors.

The purchasing power of money

Whether the purchasing power of money should be taken into account in sales forecasting depends to some extent on how it was taken care of in the analysis of demand, and to some extent on the statistics used as measures of the several factors. If the suggested preferred method of accounting for changes in the purchasing power of money in the analysis of demand were adopted, i.e., by deflating prices, personal income, advertising expenditure and any other value series by reference to the retail price index rather than counting the retail index as an independent factor in its own right, then it would seem that the resulting forecasting equation would necessitate the same treatment's being meted out to the forecast value figures of the factors.

A problem like this presents itself to the management statistician as a challenge to his ability to think, to his resourcefulness, and perhaps also to his integrity. There is one question that needs answering before a solution can be attempted. Do the planners and policy-decision makers want their forecasts in terms of undeflated money values or in terms of real money values? It is safe to say that the vast majority of them in industry and commerce would prefer undeflated money values. Forecast sales in terms of quantity would not be affected by deflation; but it is important to make sure that the forecasting procedure is consistent with the parameters of the forecasting equation. If the forecast prices etc. are left undeflated, it is likely that in these days of chronic inflation, the forecast sales in terms of quantity would be overestimated unless the calculated value of the constant K in the forecasting equation were adjusted downward proportionally to the forecast fall in the purchasing power of money.

The constant K in the forecasting equation is a pure multiplier. If it is 10 per cent higher than it should be, then the forecast value of Q, the sales by quantity, will also be 10 per cent too high. This simple fact provides a clue to the problem. It also goes to show that some deep concentrated thought is well worth while. If inflation continues at the rate it rose to early in 1970, about 10 per cent a year, it places a serious limit on the extent to which the management statistician can risk looking into the future. In times of relative economic equilibrium, when inflation does not exceed 2 or 3 per cent a year, five years is about the limit to which it is safe to go with a sales forecast. In these times of chronic disequilibrium the limit is not more than two years, which is scarcely enough for the purposes of investment and capacity planning. But for even so short a period as two years, the fall in the purchasing power of money may well exceed 20 per cent, which means that unless proper adjustments are made to the constant, K, the forecast sales could exceed the true figure by over 20 per cent.

TREND

We can now turn to the general question of trend and the methods of forecasting the trend of contributory factors. Accurate factor forecasting provides the key to accurate sales forecasting. Three questions arise:

(*a*) Does past experience of the factor reveal a trend that can be extrapolated?
(*b*) If so, is it linear or logarithmic?

(*c*) Is it worth while trying to forecast the trend of a factor by regarding it as a variable dependent on other factors and subjecting it to a regression analysis?

There is a coefficient of trend, which provides a measure of trend and its direction, and which is discussed below. Whether the trend is linear or otherwise is a matter of model building.

Of the several independent variables in an analysis of sales considered above, there is only one that could possibly be subjected to its own analysis and that is personal income. What are the factors in personal income? Are there statistics of them? Would it be any easier to forecast these factors than it would be to forecast personal income without the aid of a forecasting equation? Even if the answer to this last question is 'yes', it is questionable whether the exercise would be worth while. There could be no end to it if it is seen that every independent variable might be a dependent variable in a lower order of analysis, and that the independent variables in this lower order analysis could each be a dependent variable in a still lower order analysis, and so on.

The coefficient of trend

Usually it is obvious whether a series of statistics has an upward or downward trend or no trend at all. Sometimes it is not. The trend coefficient is designed to show the direction of trend and to provide a measure of it independently of the units of measurement used in the statistics. A measure of trend particular to any given series is the measure of its slope. For instance, the total operating costs of a company may be found to be increasing at an average annual rate of £5,000. This figure of £5,000 is a measure of trend, but expressed in terms of US dollars it would be $13,000.

The trend coefficient is independent of the unit of measurement, which means that the trend of a series in terms of tonnage or volume or French francs can be directly compared with the trend of a series in terms of US dollars or yardage or pounds sterling. However, by reference back to the series, it is possible to measure the trend in terms of its unit of measurement from the trend coefficient.

The coefficient exploits the fact that an upward trending system of weights applied to a series of figures gives a weighted average in excess of the unweighted average if the trend of the series is upward; and less than the unweighted average if the trend of the series is downward. The trend coefficient is the ratio of the weighted to the unweighted average. Where the trend is upward the coefficient is therefore greater than one; where downward, it is less than one. Where there is no trend, the weighted average equals the unweighted average and the coefficient is then one. Table 13.2 contains three examples of statistical series showing how the coefficient is calculated. The first series has an upward trend, the second series is the first in reverse, and the third according to the coefficient has no trend. The weighting system used is the simplest possible: it consists of the arithmetic progression 0, 1, 2, 3, 4,...

TABLE 13.2 The coefficient of trend

		Example I		*Example II*		*Example III*	
Period	*Weights* w	*Series* x	*Weighted series* wx	*Series* x	*Weighted series* wx	*Series* x	*Weighted series* wx
1	0	3	0	6	0	4	0
2	1	1	1	2	2	1	1
3	2	3	6	3	6	4	8
4	3	2	6	1	3	3	9
5	4	6	24	3	12	3	12
TOTALS	10	15	37	15	23	15	30
Averages:							
Unweighted (u)		3		3		3	
Weighted (w)			3·7		2·3		3·0
Trend coefficient w/u			1·2333		0·7666		1·0000

Applying the coefficient of trend

Any linear series can be extended into the future on the assumption that the coefficient of trend will remain constant. The formula for forecasting the first forecast period is:

$$F_1 = \frac{w}{2 - w/u} \qquad \ldots \text{(i)}$$

where F_1 is the trend forecast for the first forecast period, w is the weighted average, and u is the unweighted average, w/u being the trend coefficient. Applying the formula to example I, we have:

$$F_1 = \frac{3{\cdot}7}{2 - 1{\cdot}2333} = 4{\cdot}8261$$

We can confirm that this trend forecast conforms to the assumption that the trend coefficient persists into the future. The forecast of 4·8261 can be regarded as applying to period 6 of the series. Then we have:

Period	*Weights*	*Series*	*Weighted series*
1–5 totals	10	15·0000	37·00
6	5	4·8261	24·13
TOTAL	15	19·8261	61·13

Averages:

Unweighted $\frac{19{\cdot}8261}{6} = 3{\cdot}3043$

Weighted $\frac{61{\cdot}13}{51} = 4{\cdot}0754$

Trend coefficient $= \frac{4{\cdot}0754}{3{\cdot}3043} = 1{\cdot}2333$

The trend forecast of 4·8261 for the first forecast period is therefore correct, on this criterion.

Of the actual series, the period average is 3, which can be taken to be the trend norm for period 3, i.e. the middle period. The trend norm for period 6 is 4·8261, giving a trend increase in three years of 1·8261, so that the linear periodic increase is 1·8261/3=0·6087, which is added to each successive period's trend forecast to give the trend forecast for the following period:

Forecast period	*Trend forecast*
1	4·8261
2	5·4348
3	6·0435

and so on.

However, the method is not without its difficulties. It appears, indeed, to be somewhat uncertain. One would think that a constant added to or deducted from each figure of the basic series would make no difference to the answer, but in fact it does. If 10 is added to each figure of the series of example I, for instance, giving 13, 11, 13, 12, 16, the trend coefficient becomes 1.0538; and if one is deducted, it becomes 1·3500. It seems then that further research is needed to perfect the method. In the meantime, it is suggested that some measure of standardisation could be achieved by deducting from all the figures of an actual series the lowest figure in the series, or alternatively, using the squares of the deviations from the mean, which would require some complex adjustments to give the trend coefficient of the series itself.

Trend ratios and differences

The method of trend ratios lends itself to the determination of trends that are geometric in type like the trend of personal income. In effect, it provides a measure of the year-by-year proportional changes, and together with the method of trend differences, which provides a measure of the absolute changes, indicates the best form of model to use and gives an idea of the extent to which forecasts made by extrapolation can be depended upon. Table 13.3 demonstrates both methods in their application to the total personal income undeflated for changes in the purchasing power of money.

There seems little if any reason for inferring from the evidence of Table 13.3 that personal income describes a geometric (logarithmic) trend rather than an arithmetic (linear) trend. However, the evidence of a longer period of years definitely suggests that a geometric trend is appropriate. Generally, a short run of the most recent years' figures available is the best to use for determining the appropriate ratio of trend or the trend difference for forecasting purposes; but for deciding whether the trend is geometric or arithmetic a rather longer run of

TABLE 13.3 The methods of trend ratios and trend differences applied to the total UK personal income

Year	*Personal income £'000 million*	*Trend ratios*	*Trend differences £'000 million*
1964	27·7	—	—
1965	30·1	1·087	2·4
1966	32·1	1·066	2·0
1967	33·7	1·050	1·6
1968	36·2	1·074	2·5
1969	38·6	1·066	2·4
Average	33·1	1·068	2·2

years' figures is required. Government economic policy is largely to blame for the relatively low ratio and difference in 1967.

Regression trend

Chapter 7 gives the appropriate model in equation (ii) of that chapter, for applying any method of regression analysis to a practical problem of a geometric progression. Equation (i) of Chapter 3 is the actuarial version of the model. The version in Chapter 7 reads:

$$\log Y = \log b + n \log r \qquad \ldots \text{(ii)}$$

where Y is the dependent variable; b, the unattached constant; n, the independent variable representing time; and r is the common ratio. Applying least squares, we have the calculations shown in Table 13.4. The values of n for the forecasting equation follow on from the values assigned to it in the table, so that for 1970, $n=6$; for 1971, $n=7$ and so on. The forecast for 1970 is derived as follows:

$$\log Y = 1{\cdot}447 + 0{\cdot}028 \times 6$$
$$= 1{\cdot}615$$
$$\therefore \; Y = 41{\cdot}21 \text{ in £'000 million.}$$

TABLE 13.4 Least squares applied to a logarithmic time series (the total personal income)

Year	*Time* n	*Personal income £'000 million* Y	log Y	n^2	n log Y
1964	0	27·7	1·443	0	0
1965	1	30·1	1·479	1	1·479
1966	2	32·1	1·507	4	3·014
1967	3	33·7	1·528	9	4·584
1968	4	36·2	1·559	16	6·236
1969	5	38·6	1·587	25	7·935
TOTAL	15		9·103	55·0	23·248
Average	2·5		1·517		
Correction sums				37·5	22·758
Sums of squares, etc. of deviations				17·5	0·490

$$\log r = \frac{0{\cdot}490}{17{\cdot}5} = 0{\cdot}0280$$

$$\therefore \; r = 1{\cdot}067$$

Solve for log b by applying equation (ii) above to the average values of log Y and n, that is:

$$\log b = 1{\cdot}517 - 2{\cdot}5 \times 0{\cdot}0280$$
$$= 1{\cdot}447$$

The complete forecasting equation is therefore:

$$\log Y = 1{\cdot}447 + 0{\cdot}0280\,n$$

Results compared

It should be observed that the value of r of 1·067 is much the same as the common ratio of 1·068 calculated by trend ratios in Table 13.3. The common ratio calculated by the trend coefficient method applied to the logarithms is 1·062. Forecasts of the total personal income for the five years 1970 to 1974 for each of the four criteria are given in Table 13.5. The figures given are trend forecasts, and do not take account of disturbing factors, such as changes in Government economic policy and in industrial relations.

TABLE 13.5 Forecasts of the UK personal income, 1970–74, £'000 million

Year	*Trend coefficient method*	*Trend ratios method*	*Trend differences method*	*Method of least squares*
1970[1]	40·6	41·4	40·8	41·2
1971	43·1	44·2	43·0	44·0
1972	45·7	47·2	45·2	46·9
1973	48·5	50·4	47·4	50·0
1974	51·5	53·8	49·6	53·3

One of the problems that faces any trend forecaster is that some disturbing factors may have a permanent effect on future figures. This permanent effect can take either or both of two forms, viz.: it may raise or lower the common ratio of a logarithmic trend or the common difference of a linear trend, or it may cause an absolute constant increase or decrease in all future figures. Where either or both takes place,

[1] The figure for 1970 has now been published. It is £42,833 million, which exceeds any of the forecast figures in the table by an appreciable margin.

forecasts can go astray appreciably. That is why forecasting beyond the next five years is hazardous to say the least. But since long-range planning is essential to success in industry, the risk must be taken. Long-range sales forecasting provides the very foundation of investment programming and management development. But the management statistician can do no better than apply the parameters derived from the statistics of the recent past to the determination of his forecasts. The calculated errors of estimate are of limited use: they apply to the past, not the future.

14

Business ratios[1]

Business ratios, sometimes called *efficiency ratios*, have a control function, and since target ratios can be set, they also have a planning function. To be effective, they need to be compiled both systematically and frequently. A monthly routine would probably satisfy the needs of most large and medium-sized firms.

In its control function, a soundly based system of ratios helps in pinpointing areas of inefficiency, and of preventing creeping inefficiency, which is, or ought to be, one of top management's principal concerns. Ratios also provide a means of making fair comparisons. If a company has a number of establishments or factories differing in size, it would scarcely be reasonable to compare the profit or contribution of one with the others without first eliminating the size factor, which can be done by comparing the ratio of profit to size. The question, what is the measure of size? gives rise to the question of definitions generally.

DEFINITION

Most numerators and denominators are more precisely defined than *size*, which may be measured in terms of turnover, of net assets, of floor area, of numbers employed and so on. Even so, these more precisely defined entities vary appreciably from company to company. Accountants are fully aware of the wide variety of definitions that exist of turnover, profit, capital employed, labour, and value added, to name but a few of the entities that enter into business ratios either as numerators or denominators.

Definition was one of the first problems that the Centre for Interfirm

[1] The early sections of this chapter are substantially a reproduction of my article on 'Business Ratios', *The Certified Accountants' Journal*, July 1971. By courtesy of the Editor.

Comparison[1] had to tackle when the Centre was established in 1959 by the British Institute of Management and the British Productivity Council. A comparison firm by firm of, say, profit to capital, in which the definitions of profit and capital varied from firm to firm would serve only to mislead. Companies participating in the Centre's activities provide basic data in accordance with precise definitions which the Centre provides. This removes one of the two principal hazards. The other hazard, that caused by differences of size, is removed by the use of ratios. In the Centre's own words, it 'provides each participating firm with carefully selected figures (ratios and percentages) of *all* participating firms'. It should be mentioned that percentages are ratios expressed in percentage form.

COMPARISON

Ratios imply and facilitate comparison: and the question arises, comparison with what? The simple answer is standards of comparison, which stated more precisely may consist of industry averages, the ratios of other companies in the same industry, and the company's past performance. These provide standards or datum lines to work up or down from. Standards, if such they can be called, consisting of target ratios set by top management provide something to work up or down to.

To obtain industry averages, the company would need to subscribe to the Centre for Interfirm Comparison, which favours the median. The median merely restates the ratios of the middle firm in order of merit. As to the performance of other companies in the industry, it should be said at once that few are willing to co-operate in an exchange of data with their competitors. Some companies exchange data through their trade associations, the outstanding example in this country being the motor vehicle manufacturers, who exchange data through the Society of Motor Manufacturers and Traders, better known perhaps by its initials, the S.M.M.T.

Apart from the risk arising from differences of definition, there is little if any danger in an exchange of ratios. Ratios scarcely give away trade secrets, but an exchange of ratios helps competitors as well as the company itself, provided they know how to use them and are willing to devote time to the exercise.

[1] The Centre's address is that of the British Institute of Management, i.e., Management House, Parker Street, London, WC2B 5PT.

Comparison with past performance is an entirely internal matter. Nevertheless, the standardisation of definitions is still important. For the rest, the trend of a ratio over recent months gives the insight required into the company's changing efficiency, and the direction it is taking. A careful watch on the trend of the main ratios indicates at once whether or not there is an element of creeping inefficiency in the organisation.

TREND OF A RATIO

Some business ratios rise to indicate an increase in efficiency, some fall, and the remainder should remain constant each at its optimum. The ratio of earnings to capital should rise, and indeed all ratios which have earnings as the numerator should rise. So also should turnover to cost, sales to cost of sales and value added to capital employed.

Of the ratios that a management would hope to see falling are cost (or any part) to turnover and debtors to turnover; goods returned and the number of complaints to sales; and cost to capital employed.

Ratios that should tend to a predetermined optimum include the liquidity ratio (cash to turnover), stocks of materials to materials consumed; and work in progress, stocks of finished goods, and orders on hand to sales. The optimum values of such ratios are calculated by operational research on a probability basis—that, for instance, there is, say, a 95 per cent probability that the company will not run out of ready money to pay wages and bills, that it will not run out of any given material, that it will have enough finished goods to meet contract delivery dates, and so on, if it adopts the recommended ratios. Operational research methods can determine optimum values for a range of probabilities. From this information, the company's management accountant would be able to calculate the various annual costs of keeping each of the figures of cash in hand and so on, and to compare them with a view to facilitating a policy decision on whether it is worth the cost of making more certain of never running out of cash, materials, finished goods and so on, or whether it is worth the annual saving resulting from running a greater risk of running out of cash, etc. Cost and saving are related to the yield from the alternative uses of the finance sunk into liquidity, materials and finished goods on hand. There is no firm optimum value of these items in the same sense that there is at any given time a firm optimum selling price.

BUILDING A SYSTEM OF RATIOS

The main attributes of a good system of business ratios designed for control of the business are as follows:

(*a*) That the statistics required for denominators and numerators are readily available monthly;
(*b*) That the ratios in the system are all individually and collectively relevant and significant;
(*c*) That the comparative significance of each ratio is greater than that of the basic figure of the numerator from which it is derived; and,
(*d*) That, as we have seen, the definitions of the basic statistics remain constant and conform to acceptable standards.

Most systems have a primary ratio which broadly stated consists of profit to capital. The centre for Interfirm Comparison expresses it more precisely:

Operating profit/Operating assets.

Other precise definitions of the primary ratio are:

Profit before tax/share capital paid up;
Contribution/Operating assets;
Profit before tax/Shareholders' capital including accumulated retained profits;
Profit before tax/Net assets.

Of the denominators in these ratios, only net assets and operating assets are likely to vary from month to month. For the other ratios, it is questionable whether the ratio itself has any greater significance than the basic figure of the numerator. In short, then, a requirement of attribute (*c*) above is that the denominator as well as the numerator should tend to vary from month to month.

NUMERATORS AND DENOMINATORS

In a system or sub-system of business ratios, the same denominator is used throughout. Each one has a key ratio, i.e. a ratio which the other ratios add up to. Turnover or sales is a good denominator. A key ratio is profit to turnover, which is equal to the ratio of gross revenue to turnover less the ratio of total cost to turnover. By some definitions the gross revenue to turnover may be equal to unity. Nevertheless, this three-ratio system is a complete main system in its own right. Gross revenue can be broken down into a sub-system in which gross revenue to turnover is the key ratio, the several numerators consisting of the sales proceeds from each of the company's several products or services, and other current account revenue such as the sales proceeds from waste products. The same goes for total cost, in which the total cost to turnover is the key ratio of the system. The other numerators might be payroll, materials costs, heating, lighting and fuel, repairs and maintenance of buildings, plant and vehicles; rent and rates, office expenses and so on. The payroll may provide the numerator of a key ratio in a sub-sub-system. The numerators in this system might be direct wages, indirect wages in the production department, the stores, warehouse and distribution, and salaries in each of the several departments of the company.

From this complete system of ratios, it would bc possible to locate the cause or causes of a fall in the key ratio profit to turnover. The first step would be to examine the key ratios of the sub-system, and then the detailed ratios underlying the key ratio that shows a movement in the wrong direction. Attention should always be paid to the key cost ratio and its details, for it is here that creeping inefficiency is most likely to be revealed.

Although it may often be convenient to use the same denominator throughout a system of ratios, there is no reason why the same one should be used in the sub-systems as in the main system. The main system and sub-systems could each have a different denominator.

Denominators should be chosen with care. It is stated above that turnover makes a good denominator. Others that are used with success, not always in systems, are net assets, total assets and total cost, and often in sub-systems, manufacturing cost, selling cost, distribution cost, cost of sales, current liabilities, working capital, and gross profit.

BUSINESS RATIOS IN PRACTICE

Amongst the companies that have devised systems of business ratios, the mode of expression varies; some prefer percentages; others, what we might call unit ratios. A percentage is a percentage ratio, it is merely the unit ratio multipled by 100. It has been said that a ratio expressed as a percentage conveys the impression that the entity in the numerator forms part of the entity in the denominator. Accountants and statisticians know that profit does not form part of the capital employed, although the ratio of the former to the latter is often and indeed usually expressed as a percentage. If there is any danger at all that a director or any other member of top management might think otherwise, then unit ratios should be preferred to percentage ratios. But the danger is very remote. After all, active stock-market investors, both individual and institutional, know that neither the dividend per cent nor the dividend yield, which is always expressed in percentage form, implies that dividends form part of the nominal price of shares, or that they form part of the market price. Since the danger of misinterpretation is very small, the choice is a matter of convenience. It is important, however, that the chosen mode of expression should be made clear in tabulations, and should run through the whole system consistently.

Unit ratios have the advantage that they are much easier to handle in analysis. That is why mathematicians prefer them to percentage ratios. Actuaries, for instance, never use the percentage rate of interest in developing or demonstrating their compound interest formulae. School textbooks on algebra show the amount of a pound invested at, say, 5 per cent for seven years as:

$$\left(1+\frac{5}{100}\right)^7$$

whereas actuaries would show it as:

$$(1+0{\cdot}05)^7 \text{ or } 1{\cdot}05^7$$

The algebraic formula underlying this is fundamental in all actuarial formulae concerned with present values, annuities, sinking funds, and other formulae involving compound interest.

Some companies do not use an additive ratio system as described above. They prefer an eliminating system as it may be called, or a system of individual and unrelated ratios. Both of these have the disadvantage that they fail to pinpoint areas of inefficiency. They both may indicate broad fields of inefficiency, leaving the pinpointing process to an *ad hoc* enquiry, which would entail reference back to basic data. A routine additive system based on regularly collated statistics probably saves time in the long run.

An eliminating system begins with a general ratio, and follows up with a succession of ratios in which the numerator of one becomes the denominator of the next. An example in the field of plant control quoted from H. E. Betham in *The Principles and Practice of Management*, ed. E. F. L. Brech (Longmans, 1963), is as follows:

$$(i)\ \text{Machine availability} = \frac{\text{Machine available time}}{\text{Total machine time}}$$

(*ii*) Machine utilisation $= \dfrac{\text{Actual running time}}{\text{Machine available time}}$

(*iii*) Machine efficiency $= \dfrac{\text{Standard running time}}{\text{Actual running time}}$

(*iv*) Machine effective utilisation $= \dfrac{\text{Standard running time}}{\text{Machine available time}}$

In this system, there is a deviation from the principle in ratio (*iv*). Here the numerator consists of the numerator in (*i*); and the denominator, of the numerator in (*iii*). An additive system would use the numerators as shown, but the denominators in all would be the total machine time. Ratio (*iv*) would be dropped, and another ratio could be introduced in its place, that of machine time not available (that is, time lost in maintenance, repairs and preparation). The additive system lends itself better to statistical analysis than the eliminating system, although admittedly, such an advantage would not often be of any consequence.

Another custom in some companies is to compile standard ratios called basic norms, with which to compare the actual ratios. Basic norms are derived from past experience. They may be the best in recent months, or an average of recent months. Where the best ratio is an optimum, the basic norm could scarcely be anything other than the calculated optimum. Undoubtedly a basic norm is useful, provided the current trend is not overlooked as an indicator.

NORMAL SEASONAL VARIATIONS

Seasonal variations present what appears to be an intractable problem. Removing them from basic data is a simple albeit time-consuming task, which most elementary textbooks on statistics demonstrate. The problem is to decide whether to remove them or not. Experienced management statisticians are always chary of correcting basic statistics for anything: to use their own picturesque metaphor, they are afraid of throwing out the baby with the bath water. The question is, do the company's interpreters and users of business ratios want their ratios to reflect seasonal movements ? If they do, then to remove them would amount to throwing out the baby. If not, then the management statistician responsible for compiling them must either give reasons for his being unable to carry out the work, such as lack of data, or do his best with the seasonal data available.

Practice appears to vary appreciably in the way seasonal movements are dealt with. In some companies, correction is not attempted, and the users compare the current set of ratios with the set of ratios for the corresponding period of the previous year. In other companies, the ratios are based on the 12-month moving averages of the basic data. Neither of these methods can be described as satisfactory. Both iron out that very sensitivity to current events which is an essential feature of a good business ratio: they throw the baby out with the bath water. Business ratios are designed to provide an early warning system, and therefore need to reflect current trends in full. Both methods can convert a real trend in one direction to a trend in the opposite direction merely because of something that happened 12 months ago.

Not that there can be any harm in making a comparison with the corresponding month of the previous year, provided it is accepted for what it is worth and for what it means. As to 12-monthly moving averages, they seem to be far more trouble than they are worth for all the good they do. It is far more revealing to make a month-by-month comparison of recent months' ratios, and accept it as reflecting the normal seasonal movement as well as the kind of short-term movement that is being sought.

However, where it is thought desirable and necessary to remove normal seasonal variations, the text-book method can be applied effectively. This method calls for data over a period of at least three years. The data are arranged in columns year by year, each line representing a month from January to December. Each line is cross added to give a total for the month, and then divided by the number of years covered to give the average for each month.

The figures for each of the several years are then totalled, and cross totalled to give a figure which should agree with the total of the monthly totals column. The overall monthly average is then calculated from this, and the average monthly figures are deducted from it to give the deviations from the average. These deviations are the seasonal variations month by month, and then according to the sign, they are added to or deducted from the actual figure for the corresponding months, to give the seasonally corrected figures. Seasonal correction should be applied to the basic data of the numerators and denominators, not directly to the ratios themselves.

A simple example will make the method clearer. Suppose the available data cover three years from January 1968 to December 1970, and the overall monthly average is five. Suppose also that for the month of

April the figures in the three years are 10, 14 and 12, giving a total of 36 and an average for the month equal to 12, which exceeds the overall average by seven. This figure is the seasonal variation for April and its sign is negative. If the actual uncorrected figure for April 1971 is 11, the seasonally corrected figure is $11-7=4$.

If basic norms are used for purposes of comparison and the underlying basic data of the norms are known, the basic norm for each series of data can be adjusted so that it embodies the seasonal variations. In the example above, if the basic norm of the data is five then for the month of April this is increased by seven to give a seasonally corrected norm of 12.

Each series of basic data affected by seasonal movements would need to be dealt with separately. Unaffected series would be allowed to stand. Generally, but not always, it is sales that are seasonal and this would affect profits, which are highly sensitive to changes in sales and turnover no matter how caused. Wages costs vary seasonally in sympathy with changes in sales; they also vary in agriculture and horticulture directly with seasonal changes in the work cycle. Sometimes, the ratios them-

TABLE 14.1 The trend norm and variations

	Basic data					*Seasonal variations*
	1969	*1970*	*1971*	*Averages*	*Trend norm*	(5) – (4)
Month	(1)	(2)	(3)	(4)	(5)	(6)
January	1	8	6	5	8·625	+3·625
February	2	4	6	4	8·875	+4·875
March	3	7	8	6	9·125	+3·125
April	5	9	13	9	9·375	+0·375
May	9	12	15	12	9·625	−2·375
June	11	15	19	15	9·875	−5·125
July	14	14	20	16	10·125	−5·825
August	9	11	13	11	10·375	−0·625
September	8	10	12	10	10·625	+0·625
October	6	9	9	8	10·875	+2·875
November	7	9	17	11	11·125	+2·875
December	9	12	18	13	11·375	−1·625
TOTAL	84	120	156	120	120·000	0
Monthly average	7	10	13		10	
Monthly average of three years		–	–	10	–	

selves will not be affected, a seasonal correction to the numerator being offset by a seasonal correction to the denominator of like proportion.

TREND AND THE TREND NORM

Where basic data have a significant trend, as they often do, especially in growing companies, it will usually be necessary to determine the trend norm before seasonal variations can be eliminated. Where the trend is upward, there will be an upward trend during the earlier months of the year, and other things being equal, the figures for the later months of the year will appear to have a higher seasonal value than the figures for the earlier years. Since the figures for the months of this year would on the whole stand at a higher level than those for last year, it is clear that the trend variation must be taken care of as well as the seasonal variation. Table 14.1 shows how this can be done once the trend equation has been determined by regression analysis. It will be seen from the averages of the basic data, 7, 10 and 13, that the trend is linear, the complete trend equation derived from the curve fitted through the averages in column (4) being:

$$Y = 0{\cdot}25x + 8{\cdot}375$$

where Y is the average trend norm for the three years, and x represents time, i.e. January = 1, February = 2, . . . December = 12.

The trend norm figures in column (5) are based on the trend equation. Their average is 10, the same as the average in column (4); and the total of the seasonal variations in column (6) is zero. These provide a check on the method and the arithmetic.

Until the seasonal pattern changes, the seasonal variations in column (6) stand for all time; they are not affected by changes in the trend. If the basic figure for January 1972 is 9, then the seasonally corrected figure is 12·625 (i.e. 9 + 3·625), and if that for June 1972 is 22, the corrected figure is 16·875 (22 − 5·125).

USING THE TREND NORM AS THE BASIC NORM

There may be something to be said in favour of regarding the seasonally-adjusted trend norm as the basic norm. For this purpose, the averages in column (4) are the seasonally-adjusted trend norm figures applicable to the middle one of the three years (1970). For the months in succeeding years, the figure of 0·25 from the trend equation needs to be added month by month. It is equivalent to adding $12 \times 0{\cdot}25$ to the figure for a month to give the figure for the corresponding month of the following year. The seasonally adjusted figure for January 1971 is therefore $5+3=8$, and for January 1972 it is $8+3=11$. For June 1971, it is $15+3 = 18$, and for June 1972 it is $18+3 = 21$. The current basic ratio can be directly compared with the ratio based on the seasonally adjusted trend norms of the numerator and denominator to show what progress is being made. Where this procedure is adopted, the unadjusted trend norms and the seasonal variations given in columns (5) and (6) of Table 14.1 are superfluous for the purpose, though they may be useful in other directions.

Needless to say, the management statistician should carry out a periodical check of his basic data to make sure that the established seasonal patterns and trends are not changing. A routine examination once every 12 months should be enough. In any event, changes in the seasonal pattern and in the trend could scarcely reveal themselves in a shorter period.

15

Stock market statistics

From time to time most management statisticians are likely to be called upon to provide information of or to express an opinion upon the broad financial scene. Moreover, as a statistical forecaster, he will find a knowledge of current trends in the stock market to be of great value. What is the present level of interest rates? Is it likely to move upward or downward in the next year or two? The company may need new capital in the near future. Should it issue new equity stock, a debenture or another form of fixed-interest bearing security? What is the P/E ratio we hear so much about? Questions like these are almost bound to crop up in the course of the management statistician's work.

For our purpose, the stock market may be classified into five or six sectors, viz.:

(*a*) The market in Government bonds;
(*b*) The market in corporation loans;
(*c*) The market in industrial equities;
(*d*) The market in industrial fixed-interest securities;
(*e*) The market in overseas industrial equities quoted on the London Stock Exchange;
(*f*) Miscellaneous overseas loans and bonds quoted on the London Stock Exchange.

It will be enough if we take a look at (*a*), (*c*) and (*d*) of these categories.

THE GOVERNMENT BOND MARKET

Government bonds are variously known as *Government long-term debt*, *gilt-edged securities*, *gilt-edged* and *gilts*, the last being the most commonly used in the City of London. The market in gilts is examined in

Chapter 7 under the heading 'A Regression Analysis of the Gilt-edged Market', and there is little more to say.

Official summary statistics as published for each of the three classes of dated stocks should be treated with reserve. The three classes are:

Shorts (with lives up to five years),
Stocks (with lives of 5 to 15 years), and
Other dated stocks (with lives over 15 years).

The word *lives* refers not to the life at the time of issue, but to the remaining life on a given date during the lifetime of the stock. As time passes by stocks slowly move upward from one category to the next shortest and so on. For instance, a stock with six years to redemption this year is included in the 5–15 years category. Next year it will be transferred to 'shorts'. The procedure, which cannot be helped, tends to play havoc with the continuity and comparability of time series of summary statistics, especially when the stock transferred happens to be one with a low coupon rate of interest like 3 per cent or 3½ per cent, which has a tendency to stand at a higher price level relatively to gross redemption yields than stocks with a high coupon. However, the Central Statistical Office in its summary statistics published in *Financial Statistics* omits low coupon stocks and so preserves as far as possible the continuity and comparability of the several series.

THE EQUITIES MARKET

There is no royal road to a thorough understanding of the mechanism of the market in industrial equities; but a knowledge of the market ratios for the time being can take one a long way along the road.

Company accounts provide the basic data for all statistics except share prices. The number of shares issued by a company times the current price is in effect the stock market valuation of the company, as distinct from the nominal value of the shares, which is the so-called share capital appearing in the company's balance sheet, and which is equal to the number of fully paid shares times the nominal price. Discounts and premiums arising from shares issued at a price other than the nominal appear in the balance sheet as separate items. Discounts can ultimately be written off against undistributed earnings. Premiums cannot be distributed since they are not current account earnings, but they can be transferred to share capital through a scrip issue, that is, a free issue of shares to existing shareholders.

An addition to the number of shares issued through a scrip issue does not have the effect of increasing the stock market valuation of the company. A scrip issue is usually accompanied by a fall in the market price of the stock sufficient to maintain the market valuation at a constant figure for the time being. However, a scrip issue is considered to be a good sign and attracts investors; and although the market price of the stock may not recover completely for some time, it tends to begin rising again soon after the initial fall.

Stock market ratios are examined in some depth in Chapter 7 of my work on *Planning Profit Strategies* (Longman, 1971). The following is a brief resumé of the points made.

In the *Financial Times*, the share information service gives summary details of a large number of industrial stocks daily, and the FT Actuaries Share Indices, which are universally accepted as the standard for shares quoted on the London Stock Exchange, give averages by industry. The former contains the following information for each company subscribing to the service:

The name of the company and the share;
The closing market price on the last working day;
The change in price during that day;
The annual rate of dividend last paid;
The times covered;
The dividend yield;
The P/E ratio.

The nominal price is not often shown, even where space is available; but it can easily be calculated from the market price, P, the annual rate of dividend, D_c, and the dividend yield, D_y. Let P_c represent the nominal price, then:

$$P_c = P\frac{D_y}{D_c} \qquad \ldots \text{(i)}$$

The figures recorded for Lockwoods on 10 June 1971 in the issue of 11 June, under 'Food, groceries, etc.' are as follows:

Price, P, pence	100
Rate of dividend (per money unit) D_c	0·16
Times covered (E/D ratio)	3·1
Dividend yield, D_y per money unit	0·04
P/E ratio	8·0

The two dividend figures are shown in the *Financial Times* as 16 per cent and 4 per cent, which are equivalent to rates per unit of money of 0·16 and 0·04. The nominal price is not shown. Applying equation (i), we have:

$$P_c = 100\frac{0\cdot 04}{0\cdot 16} = 25 \text{ pence.}$$

A reference to a stockbroker's list shows this to be correct.

Subscripts c and y indicate the rate per money unit of nominal price and per money unit of market price respectively. Let E represent earnings in general, so that E_t is the total earnings; E_c, the earnings per money unit of nominal price; and E_y, the earnings per money unit of market price, the subscript t indicating the total. Similarly with P and D: P_t is the total stock market valuation; and P_c, the market price per money unit of nominal price; and P_y equals 1·0 or 100 per cent; and D_t is the total amount of dividends paid; D_c, the dividend rate per money unit of nominal price; and D_y, the rate per money unit of market price. Without subscripts, E, P and D represent earnings, price and dividends in general. In formulating ratios from the basic figures, care has to be taken to see that the numerator and denominator are expressed in the same terms. For instance, the times covered ratio is shown as E/D, and the price-earnings ratio as P/E; both, it will be observed, being expressed in general terms, so that we have:

$$E/D = E_t/D_t = E_c/D_c = E_y/D_y$$

and:

$$P/E = P_t/E_t = P_c/E_c = P_y/E_y$$

According to the stockbroker's list, Lockwoods Foods have a total issue of ordinary shares of 4,800,000. Their total nominal value at 25p a share is £1,200,000, and their total market price at 100p is £4,800,000. Total dividends paid were therefore 0·16 of £1,200,000 =0·04 of £4,800,000=£192,000, and earnings were £192,000×3·1= £595,200, which, of course, is an approximation with an error margin of 0·05 or 5 per cent. We can now draw up a table of basic statistics for Lockwoods shares. Table 15.1 contains the data. The two stock-market ratios are:

$$\frac{E}{D} = \frac{595\cdot 2}{192} = \frac{49\cdot 6}{16} = \frac{12\cdot 4}{4} = 3\cdot 1$$

and:

$$\frac{P}{E} = \frac{4{,}800}{595\cdot 2} = \frac{400}{49\cdot 6} = \frac{100}{12\cdot 4} = 8\cdot 06$$

The slight difference in the P/E ratio between 8·06 and the published figure of 8·0 is due to rounding. We have already seen that the total earnings figure of £595·2 is subject to an error margin of five per cent on account of the rounding of the published E/D figure of 3·1.

TABLE 15.1 Lockwoods Foods' ordinary shares on 10 June 1971

	t £'000	*c* pence per £	*y* pence per £
P	4,800·0	400·0	100·0
E	595·2	49·6	12·4
D	192·0	16·0	4·0

Notes: $P_c = £4{,}800/1{,}200$
$P_y = £4{,}800/4{,}800$
$E_c = £595{\cdot}2/1{,}200$
$E_y = £595{\cdot}2/4{,}800$
$D_c = £192/1{,}200$
$D_y = £192/4{,}800$

It will be noticed that D_y is the dividend yield, which is equal to D/P (P_y is always unity, i.e. 100 per cent, or 100p to the £). The dividend yield multiplied by the times covered ratio, E/D, gives the earnings yield corresponding to the dividend yield, i.e.:

$$\frac{D}{P} \times \frac{E}{D} = \frac{E}{P} \qquad \ldots \text{(ii)}$$

For Lockwoods it was:

$$\frac{192}{4{,}800} \times \frac{595}{192} = \frac{595}{4{,}800} = 12{\cdot}4 \text{ pence}$$

or

$$\frac{16}{400} \times \frac{49{\cdot}6}{16} = \frac{49{\cdot}6}{400} = \frac{12{\cdot}4}{4} \text{ pence}$$

Since the earning yield is the reciprocal of the P/E ratio, we have an internal consistency check on the figures published in the FT share information service: the times covered multiplied by the dividend yield per cent divided into 100 is equal to the P/E ratio:

$$\frac{100}{3{\cdot}1 \times 4} = 8{\cdot}06$$

We then have:

$$\frac{1}{E/D \times D/P} = P/E \qquad \ldots \text{(iii)}$$

What is the P/E ratio?

When the P/E ratio first became generally known in the United Kingdom in the mid-nineteen-sixties, many investors were puzzled by its implications and the high regard City editors appeared to have for it. Few, if any, financial commentators attempted to explain it; but on the whole, they made it clear, often only implicity, that the higher the P/E ratio the better the stock. By what quirk of logic, the investor may well have asked, could this be argued? Stockbrokers, bank managers and other knowledgeable people used to point to a high times covered ratio (E/D) as a good sign. Now, since a low E/D ratio contributes to a high P/E ratio, it seems that all those knowledgeable people were not so knowledgeable after all.

The fact of the matter is that in practice a high P/E ratio is usually but not always found in association with a high E/D ratio, the latter being more than offset in its effect on the P/E ratio by an extremely low dividend yield (D/P ratio). A low D/P ratio is the result of either a low dividend or a high market price, or both. It can therefore be said that to some extent, a high P/E ratio reflects the market's reaction to a high E/D (times covered) ratio.

However, there is more to the P/E ratio than its manifestation of the market's reaction to a high times-covered ratio. It also reflects the market's estimate of the growth potential of a share, that is, of the future course of its market price. If the estimate is a favourable one and the price is expected to rise, then the P/E ratio will be relatively high. How does the market come by its estimate of growth? Rumour plays no small part, but the company's press releases about its plans and prospects are an important factor. So, too, are reported changes in the top management team.

There are two points of view. First there is that of the investor who is seeking a more or less permanent and reasonably safe home for his capital. He wants a share with a high E/D ratio and a good dividend yield—a share which, in fact, would have a low P/E ratio. To him a rise or fall in the market price of his holding is merely a paper gain or loss. He buys his shares, not for any capital gain that might result from their later sale, but for their current revenue (or dividend) yield.

And secondly, there is that of the investor who is seeking to make an income out of capital gains. To him, the significance of a high *E/D* ratio is that the retained profits may be ploughed back into the business, so that the business can expand without additional equity or loan capital, and the whole of any consequential increase in profits would belong to the existing proprietors. Since the dividend yield is of no consequence to him, it may fall to a very low level as a result of a high percentage retained, and so the *P/E* ratio would tend to be on the high side.

Growth

It would seem then that the *P/E* ratio provides a measure of the market's esteem for a share in respect of potential capital gains. Of the two factors in the ratio, the denominator, earnings, lies outside market influences. It is the price the market determines, and when the *P/E* ratio is high, the price too is high relative to earnings. The question that arises is whether for any given share at any given time, the price of the share fully discounts potential growth. We know from the records that this cannot be so. The general upward and downward movements of the market as reflected in the share prices provide enough evidence. If growth is fully discounted when the price is at its peak, it is less than fully discounted when the price is low. If it is fully discounted when the price is low, it is more than fully discounted when the price is high, and then the share is overvalued in the market.

Since total industrial earnings remain fairly constant over the years, changes in the average *P/E* ratio are largely due to changes in price in the same direction. Any change in the *P/E* ratio due to a change in earnings would be in the opposite direction. Table 15.2 gives the averages of the years 1965–70 for industrial shares. An examination of the percentage changes in price and the *P/E* ratio indicates that almost the whole of the increase in the latter between 1966 and 1968 was accounted for by a rise in price, whereas between 1968 and 1970, rather less than half the fall in the *P/E* ratio was accounted for by price:

Percentage changes	*Price*	*P/E ratio*
1966–68	+50·9	+56·4
1968–70	−12·4	−28·3

This means that between 1966 and 1968, earnings remained fairly constant, whereas between 1968 and 1970 they increased appreciably.

Consistent with this latter is the appreciable rise in the *E/D* ratio from 1·33 to 1·52 between 1968 and 1970. Column (6) of the table gives an index of total earnings for the shares covered by the statistics. It is arrived at by multiplying *E/P* by *P* (=*E*). It should be noted that the figures have a comparative significance only, and no absolute significance at all. They show little change between 1966 and 1968, but an appreciable increase between 1968 and 1970; thus confirming the conclusions reached above.

TABLE 15.2 *Financial Times*—Actuaries index numbers, 500 industrial shares

	Price *P* (1)	*Dividend yield %* *D/P ratio* (2)	*Earnings yield %* *E/P ratio* (3)	*Times covered* *E/D ratio* (3)÷(2) (4)	*P/E ratio* $\frac{100}{(3)}$ (5)	*Index of earnings* (3)×(1) (6)
1965	106·7	5·54	7·96	1·44	12·56	849
1966	107·6	5·67	7·69	1·36	13·00	827
1967	114·9	5·16	6·89	1·34	14·51	792
1968	162·4	3·69	4·92	1·33	20·33	799
1969	160·5	3·90	5·77	1·48	17·33	926
1970	142·2	4·52	6·86	1·52	14·58	975

Source of cols. (1)–(3), *Financial Statistics*

One would expect to see a rise in the market prices of shares when earnings are increasing; but between 1968 and 1970, the opposite happened. In the period, earnings increased by 22·1 per cent, whereas the *E/D* ratio increased by only 14·3 per cent, which indicates that dividends also rose between the two years. In short, then, an explanation for the decline in the equity market that has taken place since the end of January 1969, when it reached its peak, cannot be found in the economics of the situation.

In the long run, market growth depends largely upon industrial growth, but in the relatively short period of two years prior to January 1969, the market had undoubtedly outpaced industrial growth. Investor confidence had over-reached itself, and the following two years witnessed a readjustment of the market to a level more in keeping with industrial growth.

But who is to say what the true capital value of industry may be at any time? Only the equity market can decide, and its decision is self-justified. The market changes its mind about individual stocks minute by minute during the business hours of the Stock Exchanges, and sometimes outside business hours as well. It also frequently changes its mind about stocks as a whole. Superimposed upon the day-by-day fluctuations is an upward or downward general trend, which may take months to discern.

Nominal prices

It is sometimes argued that the nominal price of a share has no meaning and should therefore be abolished as superfluous. We are not concerned here with the financial accountant's attitude to the proposal. He would have no difficulty with his balance sheet figures; he would enter the capital receipts from the issue of equities at the issue price, and since there would be no question of premium or discount, there would be no separate entry: the whole of the share capital recorded in the books of account would be the actual capital amount received, gross or net of the expenses of issue.

From the stock market point of view, there would similarly be little if any difficulty. The dividend would be shown as an amount per share instead of a percentage of the nominal price. There would be no need for any internal consistency between the nominal and market prices on the one hand and the dividend or nominal and market prices on the other. The two important ratios (E/D and P/E) would still be available.

However, the proposal would necessitate new legislation: the Companies Acts as they stand assume a nominal price and indeed the Acts have much to say on the issue of share capital. A company may, for instance, issue shares at a discount provided that the issue is authorised by resolution passed in general meeting and is sanctioned by the Court; that the shares are of a class already issued; and that the company shall have been in business for at least a year. There are rules, too, governing the disposition of the amount of a premium received on shares issued at a premium; it forms part of the capital reserve and cannot be distributed as a dividend; but it can be used in paying up shares to be issued as fully paid bonus shares to existing shareholders. Whether or not a nominal price is explicitly required by the Acts, it is certainly implicitly required by such provisions as these.

INDUSTRIAL DEBENTURES

Debenture stocks have recorded a steady fall in their price level over the years. The *Financial Times*—Actuaries price index of redeemable debenture and loan stocks with 20 years to maturity records a fall from 99·35 in 1965 to 70·46 in 1970, the redemption yield rising from 7·07 per cent to 10·50 per cent over the same period. This means in effect that few companies could hope to issue a new debenture at par redeemable in 20 years or over for less than 10 per cent, or a new irredeemable unsecured debenture for much less than 12 per cent. Length of life plays a part in the yield of industrial debentures as it does in the yield of Government and local authorities' bonds. On the whole, it seems that the most creditworthy companies must pay in interest at least $1\frac{1}{2}$ per cent more than the Government for stock of the same length of life.

Whether interest rates had reached their peak by March 1971, when 20-year industrial debentures stood at an average yield of 10·68 is questionable. The principal factors are the same as those influencing the level of Government bond yields, by far the most important being the rate of inflation. Until inflation is contained by a cessation of wage settlements representing a rate of increase in wages greater than the rate of increase in productivity, there can be no hope of a reduction in interest rates to the level prevailing in the early nineteen-fifties.

Oddly enough, the yield of preference stock always stood on average at a lower level than the redemption yield of 20-year debentures up to September 1970. Then in October, it averaged 10·34 per cent compared with 10·27 per cent for 20-year debentures. By March 1971, the new formed gap had widened with averages of 10·97 per cent and 10·68 per cent; it narrowed slightly between March and June 1971, the figures for the latter date being 10·62 per cent and 10·37 per cent respectively. By mid-September 1971, the gap had widened to 0·33 per cent, with debentures at 9·62 per cent.

CAPITAL GEARING

Capital gearing, sometimes called leverage, is a concept based on the theory that the higher the ratio of loan capital to share capital, the greater is the risk attached to equity income. Loan interest forms part of the gross income, and the greater the proportion of gross income consisting of loan interest, the smaller is the amount of income for the shareholders. But that is not all. Loan interest is a cost and must be deducted from gross income to determine net income. Every change in gross income results in a corresponding absolute change in net income. If the gross income of a company is £1,000 in 1970, and falls to £750 in 1971, the amount of loan interest in both years being £500, then although the gross income falls by 25 per cent, the net income falls by 50 per cent from £500 to £250.

The *Financial Times* share information service gives no information about capital gearing; but the gearing of individual companies' financial structures is often mentioned in City editors' comments. Probably more importance is attached to the capital gearing ratio than should be. After all, loan interest is a current cost like any other current cost. And like other current costs it is a deductable expense for corporation tax purposes.

Unquestionably the taxation aspect provides an encouragement to companies to obtain new finance by issuing new loan capital rather than new equity stock; but this factor is probably more than offset by the prevailing rates of interest payable on loan stock.

There is another aspect. Academics tend to divide capital sources into two distinct markets, viz.: the market in equity stock and the market in loan stock. However, in practice the division is not so clear cut. There are voting and non-voting shares, which may not be of any importance in the division of the capital market. But preference shares and cumulative preference shares, which are fixed interest bearing securities in effect, appeal to a section of the market more closely related from the point of view of taxation to the ordinary shares market than to the loan market, but more closely related from investors' point of view to the loan market than to the shares market.

Debenture holders take priority over preference shareholders, and preference shareholders take priority over ordinary shareholders. This represents the order of risk, so that it seems odd that the yield on preference shares consistently stood at a level below that of debenture stock for some years prior to the end of 1970. There may be technical factors that explain it, such for instance, that the market price of preference shares may tend to rise with industrial growth whereas the price level of debentures is constantly under the pressure exerted by inflation.

Another factor that tends to weaken the capital gearing ratio as an investors' tool is that it is necessarily based on nominal or historical valuations of equity and loan capital rather than upon current market valuations.

16

Decision trees

Decision trees are an academic concept of great potential practical value. Their main branches represent the alternative decisions that can be made and the secondary branches provide subjective judgments dressed in objective clothing. Their greatest value lies in capacity planning; but they are nothing if not versatile. Wherever there is a decision to be made, there are alternatives, and whenever there are alternatives, there is room for a decision tree or a simple adaptation of a decision tree.

Consider the example of a flour confectioner who finds it convenient to purchase all his doughnuts from another bakery. He buys them by the dozen and makes 50 per cent profit on his selling price of 20p a dozen. His problem is to decide how many to order each day. He reckons there is a 20 per cent chance of selling four dozen a day: a 50 per cent chance of selling three dozen a day, and a 30 per cent chance of selling two dozen a day. How many should he order to maximise his profit? In the past, he might have argued: 'If I order two dozen a day I will probably sell all of them, and there will be no waste. If I order three dozen, I might not be able to sell them all on about one day in three, and there would be considerable waste. And if I buy four dozen, there would be waste on four days out of five.'

However, the decision tree is concerned not with minimising waste, but with maximising profit. A tree for the confectionery example in the form as usually expounded is given in Fig. 16.1. It takes root at the box marked D, and gives off three main branches, each representing one of the ordering options open to the confectioner. Then each main branch has three secondary branches, each providing for an estimate of sales and the resulting profit.

Consider the top main branch representing the purchase of four dozen doughnuts a day. The daily cost will be 40p no matter how many

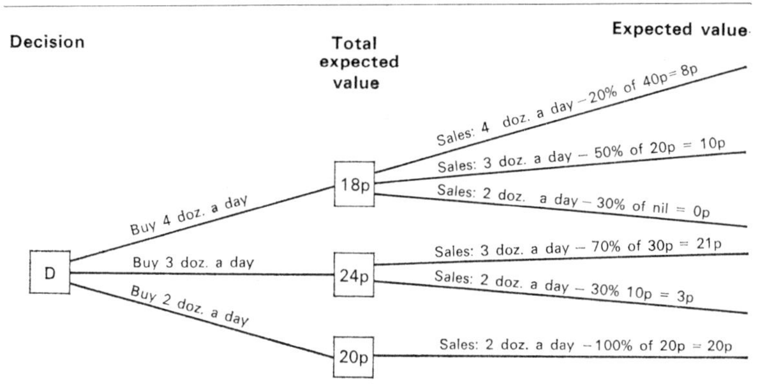

FIG. 16.1 The decision tree for doughnuts

are sold. On 20 per cent of days all four dozen will be sold, so that for those days the gross receipts are four dozen at 20p a dozen equals 80p a day, and the gross profit is 80 – 40 = 40p. On 50 per cent of days, sales will be three dozen giving gross receipts of 60p a day, which, less the cost amounting to 40p, yields a daily gross profit of 20p. On 30 per cent of days, sales will be two dozen giving gross receipts of 40p, which is just equal to the cost, so that the profit on those days is zero.

Taking into account the percentages of days, we have the following for a daily purchase of four dozen:

Daily sales in dozens (1)	*% of days* (2)	*Daily profit* (p) (3)	*(2) of (3)* (p) (4)
4	20	40	8
3	50	20	10
2	30	0	0
TOTAL—the expected daily value			18

The logic of the decision tree can be demonstrated by a somewhat more practical and less esoteric approach.

A LESS ESOTERIC APPROACH

This removes the esoteric flavour from the academic demonstration. It is to calculate the total profit earned in a period of 100 days:

Daily sales in dozens (1)	No. of days (2)	Daily profit (p) (3)	Total profit (2) × (3) (4)—(p)
4	20	40	800
3	50	20	1,000
2	30	0	0
Total	100		1,800
	Daily average		18

It gives the same daily profit as the academic demonstration.

For the second main branch, the one on which the order is three dozen daily, it will be seen that as it is not possible to sell more doughnuts on any day than the number ordered on that day, the 20 per cent for four dozen can be merged with the 50 per cent for three dozen, so that, in effect, there are only two secondary branches. Here, the total daily profit is 3 × 10p = 30p and sales and profits are:

For three dozen: sales proceeds = 60p; profits 30p a day.

For two dozen: sales proceeds = 40p; profits 10p a day.

Using the less esoteric 100 days approach, then, we have:

Daily sales in dozens	No. of days	Daily profit (p)	Total profit (p)
3	70	30	2,100
2	30	10	300
Total	100		2,400
	Daily average		24

For the third main branch there is, in effect, only one secondary branch, the daily cost being 20p, sales proceeds 40p and the daily profit 20p.

100 days at 20p profit a day = 2,000p.

A summary of the results may be stated as follows:

Size of daily order	*Profit in 100 days* (£)	*Average daily profit* (p)
Four dozen	18	18
Three dozen	24	24
Two dozen	20	20

On the assumptions underlying the decision tree of Fig. 16.1, the most profitable size of daily order is three dozen, and the next most profitable is two dozen.

THE TWIGS OF THE TREE

It seems that the decision tree as expounded academically imposes an unrealistic constraint on the sales side of the business. That doughnuts are bought and sold at wholesale in units of a dozen seems realistic enough. But that does not mean that a similar constraint must be imposed on sales at retail. When it is stated that four dozen doughnuts are sold on each of 20 days out of 100, it means that the demand for doughnuts on 20 days out of 100 stands at four dozen or more. But when four dozen a day are purchased, no more than four dozen are available to satisfy the demand.

Similarly, when three dozen a day are purchased, only three dozen are available, although it is known that on 20 days out of 100, four dozen or more could be sold. There is reason for supposing that on some of the 50 days more than three dozen but less than four dozen could be sold. The academic decision tree sweeps these into the three dozen fold as though they counted for nothing more. Perhaps they do on some occasions; but nobody can be certain until the necessary estimates and calculations have been made.

In a case like this, the percentages need not be subjective as implied above. Suppose the flour confectioner has had two or three years' experience selling doughnuts from a daily order of four dozen. Then his estimated percentages of sales could be based on a carefully kept daily record of actual sales. Suppose he makes a complete statistical analysis of the different levels of actual sales showing the average number of days out of 100 for each level to the nearest quarter of a dozen. The figures could very well look something like those in Table 16.1, which consists of what might be called the tabular decision tree complete with twigs. The distribution is near normal, the daily sale occurring most frequently being 36, apart from 48, which has the constraint of being the maximum that can be sold in a single day, and therefore covers potential sales of upwards of four dozen daily.

Putting in the twigs of the tree changes the picture somewhat. The three dozen order still ranks as the most profitable; but now the four dozen order runs it a fairly close second, and the two dozen order falls to a poor last by comparison.

TABLE 16.1 A tabular decision tree with twigs

			Daily rates			
Sales daily No (1)	*Frequency of sales* Days (2)		*Sales proceeds at 20p a doz* (p) (3)	*Profit* (p) (4)	*Total profit in 100 days* (2) × (4) (p) (5)	(p) (6)
Daily order: Four dozen: . . . (3)—40p						
48	20	20	80	40	800	
45	9		75	35	315	
42	12		70	30	360	
39	14		65	25	350	
36	15	70	60	20	300	
33	14		55	15	210	
30	8		50	10	80	
27	5		45	5	25	
24	3	100	40	0	0	2,440
Daily order: Three dozen: . . (3)—30p.						
36	70		60	30	2,100	
33	14		55	25	350	
30	8		50	20	160	
27	5		45	15	75	
24	3	100	40	10	30	2,715
Daily order: Two dozen: . . (3)—20p.						
24	100	100	40	20	2,000	2,000

It looks as though a diagram of the decision tree such as that depicted in Fig. 16.1 is quite superfluous. It is the academic picture on the box: if there is anything in the old proverb 'Don't judge the contents by the picture on the box', the tree is probably best ignored. The whole process can be, and is probably best done in a statistical tabulation of the kind presented in Table 16.1. The use of percentages and the words *chance* and *probability* in the demonstrations commonly found in the literature, at schools of business studies, and at management seminars, gives the concept an esoteric mathematical flavour which many people experience difficulty in stomaching.

The use of a simple frequency distribution with numbers of days or other units of time instead of percentages, and allowing the tree to sprout its natural twigs, not only eliminate the esoteric and pseudo-mathematical flavour, they also make the whole concept of decision analysis much easier to understand, not to mention the more accurate solution that results from them.

AN EXAMPLE FROM REAL LIFE

An article in *The Financial Times* of 18 May 1971 gave details of a decision analysis of a problem with a complication. The decision tree given in the article was used by Len Sneddon, director of Sundridge Park Management Centre for a firm of domestic appliance manufacturers. It concerned a new type of gas fire and water heater. The problem was to decide whether the firm should go on producing 80 units a week, or cut down to 60 units. The firm had 520 units in stock, enough to supply 20 units a week for six months, which was the forecasting period; so that with an output of 80 units a week, the firm would be able to supply a total of 100 units a week for six months.

The firm estimated that there was a 25 per cent chance that sales would rise to 100 a week, a 60 per cent chance that they would remain at 80 a week, and a 15 per cent chance that they would fall to 60 a week. Translated into terms of weeks out of 26, the percentage figures give 6·5 weeks, 15·6 weeks and 3·9 weeks respectively. Before the method of Table 16.1 can be applied to the problem, it is necessary to deduce, from the data given in the article, two or three items. They are the selling price, which is £100 and the marginal cost, which turns out to be £40·40. It seems also that the firm took into account a fixed cost amounting to £1,960 a week. Table 16.2 sets out the calculations, in which sales can exceed output by 20 units a week. The cost figures exclude the cost incurred in producing the stock of 520 units, unless it accounts in part for the apparent fixed cost of £1,960 at a marginal cost of £37·69.

The total expected values of £78,200 and £86,200 given in the last column agree with Mr Sneddon's figures. They indicate that other considerations apart, it would pay to reduce output to 60 units a week.

TABLE 16.2 A tabular decision tree without twigs for the gas heaters example

Weekly sales No. (1)	*Frequency in weeks* No. (2)	*Weekly rates. Sales proceeds at £100 a unit* (£) (3)	*Profit* (£) (4)	*Profit in 26 weeks* (2) × (4) (£) (5)	*Total expected value* (£) (6)
Weekly make: 80 units:			*(3) – £5,192*		
100	6·5	10,000	4,808	31,250	
80	15·6	8,000	2,808	43,800	
60	3·9	6,000	808	3,150	78,200
Weekly make: 60 units:			*(3) – £4,384*		
80	22·1	8,000	3,616	79,900	
60	3·9	6,000	1,616	6,300	86,200

TABLE 16.3 A short-cut method

Weekly sales No.	*No. of weeks*	*Total sales* No.	(£)
100	6·5	650	
80	15·6	1,248	
60	3·9	234	
	TOTAL	2,132	
Sales proceeds at £100 =			213,200
Total cost of 80 units a week			135,000
Total expected value			78,200
80	22·1	1,768	
60	3·9	234	
	Total	2,002	
Sales proceeds			200,200
Total cost of 60 units a week =			114,000
Total expected value			86,200

Another method of arriving at the same solution is somewhat shorter: calculate the total number sold in 26 weeks, charge out at the price, and deduct the cost incurred in 26 weeks; as in Table 16.3.

There is reason for inferring that the exercise was entirely concerned with the rate of production, and that the market was expected to remain constant. If the firm could sell 2,132 units when 80 a week are made, it could sell the whole output and the stock when 60 a week are made up to a limit of 2,132 units. The total production in 26 weeks at 60 a week is 1,560, which together with the stock of 520 units, makes a total of 2,080. This at £100 a unit less the cost, gives a total expected value of £94,000 when 60 units a week are produced.

The implication of this is that in some weeks sales would exceed 80 units, and would reach 100 on occasions until the stock of 520 had run out, when the rate of sales would actually be constrained by the rate of output.

There does not appear to be any good reason why the rate of output and the available stocks should impose a constraint on the rate of sales. It is unrealistic to suppose that there would be exactly 60, 80 or 100 sold each week during the 26 weeks. There would be some weeks in which 61 or 62 or any other number up to about 110 or so would be

TABLE 16.4 A tabular decision tree with twigs for the gas heaters example

Weekly sales No.	*No. of weeks*	*Total sales in 6 months* No.	*Sales proceeds @ £100 a unit* (£)	*Cost* (£)	*Expected value* (£)
Weekly make: 80 units					
110 and over	1·5	165			
105–109	2·0	210			
100–104	3·0	300			
95–99	4·0	380			
90–94	4·5	405			
85–89	4·0	340			
80–84	3·1	248			
75–79	2·0	150			
70–74	1·0	70			
65–69	0·6	39			
60–64	0·3	18			
Total	26·0	2,325	232,500	135,000	£97,500
Weekly make: 60 units:					
Total	26·0	2,080	208,000	114,000	£94,000

sold. In short then, it seems that the figure of 60 units sold during 3·9 weeks really means 60–79 units, the figure of 80 means 80–99, and that of 100 means 100 and more. What is needed is an estimate of the number of days on which each number from 60 upwards would be sold. With such estimates, the whole picture might change.

Table 16.4 gives an estimated spread of the numbers of days in steps of five units at a time, and the calculations by the short-cut method. Of the weekly sales the lowest figure has been used for calculating the total sales in six months. If the middle figure in each range had been used, the total sales in 26 weeks would have appeared as about 2,347, well within the total make plus the opening stock totalling 2,600 units, and the expected value would be £99,700 instead of £97,500 as shown.

Seen in this light, with the removal of the production rate constraint on estimated sales, there is no doubt that the most profitable rate of production was 80 units a week, rising to 100 units a week when stocks on hand had been run down. For the estimated level of demand for the gas heaters, a production rate of 100 a week would lead to an accumulation of unsold stocks on hand of some 250 units every six months, with a variable cost at £40 each of £10,000. A production rate of 80 units a week would on the evidence fall short of demand in six months by a total of about 240 units, which would mean a loss of profit on variable cost at £60 a unit (£100−£40) of £14,400.

Perhaps the old fashioned way of doing an exercise of this kind provides the best method. It goes unsung, and it has no academic picture on the box. The problem is whether, when stocks have run out, it would pay to produce 80 units a week or 100.

	£
Output: 80 units a week:	
Sales in six months 2,080 units @ £100 a unit	208,000
Cost of 2,080 units	135,000
Profit	73,000
Output: 100 units a week:	
Sales in six months 2,325 units @ £100 a unit	232,500
Cost of 2,600 units made	156,000
Profit	76,500

There is a possibility that an increase in the rate of output to 100 units a week would involve the company in additional fixed costs, consisting for instance, of annual capital charges on additional fixed assets required to increase the capacity of the plant. This possibility is ignored in the above calculations.

Pricing policy and plant capacity

The accumulation of unsold stocks of 500 units a year when output stands at the rate of 100 units a week gives rise to the question whether it would pay to reduce the price in order to dispose of them. We have no knowledge of the optimum price or of the elasticity of demand. We have the actual price, £100, and the marginal cost, £40·4 (say £40), and from these we can calculate the elasticity of demand implied by the price, and consider whether it is reasonable or not. The formula for the implied elasticity of demand, e_i, can be derived from the optimum price formula. It is:

$$e_i = \frac{p}{p-a}$$

Substituting, we have:

$$e_i = \frac{100}{100-40} = \frac{100}{60} = 1{\cdot}667$$

Experience indicates that for items of consumer durables like gas heaters, the elasticity of demand is relatively high, and in this case, may be as high as two. In any event, there is not much doubt that the figure of 1·667 is on the low side. If the correct figure were two, then the optimum price would be £80:

$$p_m = \frac{\pounds 40 \times 2}{2-1} = \pounds 80$$

and if it were 1·8, the optimum price would be £90:

$$p_m = \frac{\pounds 40 \times 1{\cdot}8}{1{\cdot}8-1}$$

We may calculate the effect on profits of reducing the price to an optimum of £80 by first calculating the value of k in equation (vi) of Chapter 8. From this equation, we have:

$$k = Qp^e$$

For $Q = 2{,}325$ in six months, $p = \pounds 100$, and $e = 2$, we have:

$$k = 2{,}325 \times 100^2$$
$$= 23{,}250{,}000$$

For $p = £80$, then, using equation (vi) Chapter 8, we have:

$$Q = \frac{23{,}250{,}000}{80^2}$$

$$= 3{,}633 \text{ units in six months}$$

which at £80 a unit gives total receipts of £290,640. Imposing the same constraint on the rate of production, that it can be increased or decreased only in steps of 20 a week, we find that the firm would have to make a total of 140 a week or 3,640 in six months to meet the demand for 3,633 at £80 a unit. Assuming there is no additional fixed cost, then we have the following position:

Output: 140 units a week:

	£
Sales in six months: 3,633 units at £80 each =	290,640
Cost of producing 3,640 units =	197,600
Profit in six months =	93,040

If an output of 140 a week called for a plant expansion of one unit of capacity at £200 fixed cost a week, the profit of £93,040 would fall to £87,840.

CONCLUSION

Decision trees and their more practical variants have a wide application in the planning of capital expansion projects. They can be used in ranking alternative projects and in estimating the optimum capacity of plant required, as well as in ordering goods.

It will be seen in Table 16.4 that a key estimate of weekly sales is the mode, i.e., the peak of the frequency distribution, viz: 90–94 units weekly. In forming a frequency distribution, the management statistician would find an estimate of the mode and its frequency of great value. When the market research office is asked to give its forecast of weekly or monthly sales of a product, it should be invited to give the estimated mode outside the pattern of the forecast sales. It will be seen that the forecast weekly sales of 100 (25 per cent), 80 (60 per cent), and 60 (15 per cent) account for 100 per cent, the assumption being that they will never in any week fall below 60. An estimate of the mode and its frequency necessarily lies outside this pattern.

It should be mentioned that the distribution given in Table 16.4 has a slight right-hand skew, the arithmetic mean of weekly sales being

89·5, which is somewhat lower than the mode. No significance should be attached to this: it is due to a squeezing into one group of the frequencies shown for sales of 110 and upward a week.

One pitfall to avoid is allowing the organisational constraints on the rates of output to dictate the pattern which forecast sales must take. With durable goods, which can be drawn from or put into stock, there is no need to set an upper limit to the rate of sales. In the long run, of course, no more can be sold than are made.

The problem calls to mind the bewilderment caused by the monthly statistics of production and sales of domestic television sets published in the *Monthly Digest of Statistics* just before the Coronation of Queen Elizabeth II. The figures showed that in the previous six months, sales had far exceeded production. A glance at the figures for earlier months provided the solution to the mystery: production had then far exceeded sales. The manufacturers had been making for stock in anticipation of a greatly increased demand for the Coronation.

17

Stock control

Materials stock control is a sphere in which the management statistician may play a leading role. Statistics lie at the centre of any control system. There are three basic stock statistics, viz: (*a*) inflow, (*b*) level of stock, and (*c*) outflow. In the long run inflow must be great enough to meet the outflow; and the level of stock must be high enough to provide a buffer against any variation in outflow that may not easily be met by changes in the rate of inflow.

Since holding stocks of materials is expensive, there is a tendency to keep them to a minimum. Food manufacturers are compelled by the perishable nature of many of their materials to work on a hand to mouth basis. To them, regular daily deliveries of eggs, milk, butchers' meat and soft fruits are almost essential. Buffer stocks, especially of milk, are out of the question. Some large engineering concerns insist upon daily deliveries from their suppliers; but this does not mean that they do not maintain buffer stocks.

Material stocks are thus subject to two diametrically opposed pressures, viz: the downward pressure exerted by the financial burden they impose upon the company, and the upward pressure exerted by the need to maintain stocks at a level high enough to meet the requirements of the production or sales department. Whatever the level of stocks may be, there is always a risk of disruption in the production department on account of a shortage of materials. Industrial action by the employees of the suppliers, by transport men, or by the company's own reception-bay men is a factor that manufacturers can scarcely afford to insure against by maintaining high stock levels.

Materials that have to mature, like potable spirits, or that have to season, like timber, are necessarily held in large quantities, and whether they are held by the producer or the ultimate bottler or manufacturer, they have to be financed and accommodated in warehouse or shed; and the cost has to be carried in the price to the ultimate consumer.

In these days of high interest rates and the cost of capital, the downward pressure exerted by the financial burden is so much the greater, and the tendency now is to consider to what extent the burden can be reduced by ordering policy. It seems the best way of minimising stocks by ordering policy is to order less at more frequent intervals.

Needless to say, a small order policy is not to the liking of suppliers, who find themselves carrying a greater stock burden in consequence. They are countering the tendency by increasing the differential discount for large deliveries, called, in the jargon, *the vendor set-up cost*; and by raising a fixed charge per delivery, called the *vendor set-up charge.* In the literature, the word *order* is often used in place of *delivery*. The latter is preferable if only because it accommodates standing orders the better. A standing order to a supplier for a monthly delivery of 100 tons of material is more economical to him than an order for a weekly delivery of, say, 23 tons.

THE ECONOMIC ORDER QUANTITY

The economic order quantity, EOQ, sometimes referred to as the economic purchase quantity, EPQ, is the optimum size of order for any given material or bought out part; that is, it is the most economical size of order from the purchaser's point of view. The expression is also used of orders placed on the production department for supplies of intermediate products, some of which may be sold as spare parts.

A formula, with variants, for EOQ has been bandied about for some years. Its origin is obscure; but it was probably devised by operational research. Its logic, too, is far from clear. It expresses EOQ as a function of various factors, some of which are mentioned above. The factors are grouped as follows:

1. Stock carrying cost, covering:
   (*a*) Risk of obsolescence and deterioration in store;
   (*b*) Insurance of stocks, warehouses, and handling plant against loss by fire etc.;
   (*c*) Interest on value of stock; and
   (*d*) Interest, depreciation, repairs maintenance and labour cost of storage and handling;
2. Ordering cost, including:
   (*a*) Internal cost of ordering;
   (*b*) Cost of reception etc.; and
   (*c*) Vendor set-up cost and charge;

3. Annual usage;
4. Vendor's quoted price.

It seems that the standard formula is:

$$E=\sqrt{\frac{2UB}{Ap}} \qquad \ldots \text{(i)}$$

where E represents the EOQ; U, the annual usage; B, the ordering cost; A, the annual stock carrying cost expressed as a percentage of the average stock value; and p, the supplier's quoted price. In order to strip the formula of the esoteric academic flavour given to it by the square-root sign, we can restate it:

$$E^2=\frac{2UB}{Ap} \qquad \ldots \text{(ii)}$$

How is the formula derived? In the literature, it is quoted as though it were a self-evident truth or a piece of inescapable dogma which we must accept without question just as a good Roman Catholic believes in the Immaculate Conception. This is not to say that the formula is arbitrary, that nobody ever set out a full mathematical proof. But let us examine it in the light of the factor definitions. Most writers are satisfied to give very sketchy definitions. For instance, the most elaborate definition given above is that of the annual stock carrying cost, which is expressed as a percentage of the average stock for the purpose of the formula. One writer defines it for the same purpose, in the following terms: 'the interest + storage charge per unit per unit time'; and another, 'the incremental cost of holding items in stock', such variations in definition cannot be ascribed to a need for brevity: they are too fundamental for that. The phrases *percentage of average stock*, *per unit per unit time*, and *incremental* have entirely different meanings. The inconsistencies call to mind the way a story changes as it is passed on by word of mouth; and they are disconcerting to say the least.

It is worth mentioning that the detailed definitions given above are based on those stated in an article on 'Economic Purchase Quantity Calculations' by J. R. Rhinehord in *Management Accounting* of the USA, September 1970. There is reason for supposing that they conform more closely to the original definitions than the less detailed ones given by other writers.

In definitions of this kind expressed for factors in a formula, there is always a danger that some may beg the question in the sense that a solution is required before a factor can be evaluated. It seems that one such factor in the formula of equations (i) and (ii) above is ordering cost, which includes the vendor set-up cost, i.e., the price variation for size of order. How can one take into account the variation from the price quoted until one knows the size of the order?

Factors included in the numerator of the formula would have the effect of increasing the solution value of E, and those included in the denominator, would have the effect of decreasing the solution value of E. Let us examine the factors in this light.

Factors of the numerator:

U, *the annual usage.* Undoubtedly, the greater the annual usage, the greater would be the solution value of E.
B, *the ordering cost.* Again there is no doubt that the greater the ordering cost per order, the greater the value of E. The difficulty with this is that it holds true only to the extent that the ordering cost varies with the number of orders placed. The vendor set-up charge is strictly variable; but it is questionable whether the internal cost is anything but a fixed or time cost in most stores departments.

Factors of the denominator:

A, *the annual stock-carrying cost.* The greater the costs that vary directly as the stock held (interest on capital sunk in stocks, insurance of stocks) the smaller the optimum stock, and therefore the smaller the value of E. But most of the other stock-carrying costs are non-variable time costs which can scarcely affect the value of E: warehouses and plant do not expand or contract as stocks increase or decrease. Even stores handling labour is not likely to vary fully with stocks held.
p, *supplier's quoted price.* The higher the price, the greater is the value of the stock and therefore the variable carrying cost; so that price belongs to the denominator.

There are two factors, both belonging to the numerator, which the formula of equations (i) and (ii) does not take into account. One is seasonal variations in usage, and the other, which is of great importance nowadays, is inflation. The greater the seasonal variations, the greater a buffer stock needs to be; and where the rate of inflation is high it may pay to buy materials now for stockpiling, even at current rates of interest.

The figure 2 in the numerator may be regarded as a coefficient. Some writers who have monthly average usage instead of the annual usage in the numerator, have 24 instead of 2. Apart from the apparent inclusion of a question-begging factor and a failure to specify variable costs, the formula cannot be faulted. But this is purely negative evidence; it is no proof that the formula is valid.

A numerical example

One way of testing the validity of a mathematical formula is to apply it to a numerical example. It is not enough to show that it gives a reasonable solution by subjective judgment. In an optimising formula, it must be shown to give the optimum solution.

Suppose a road haulage contractor wishes to optimise his orders for diesel fuel oil. The firm's annual consumption is 30,000 gallons; the quoted price, 20p a gallon for deliveries of up to 1,000 gallons. There is a discount of 1p a gallon for deliveries of 1,000–5,000 gallons, and of 2p a gallon for 5,000 gallons and over. The vendor set-up charge is £2·00 per delivery. There is no risk of obsolescence or of deterioration in store, and the internal ordering costs and storage costs are entirely non-variable. The contractor maintains a buffer stock of 1,000 gallons, enough for about 10 days' consumption, and the maximum stock depends on the size of the delivery. If he orders 10,000 gallons at a time, the maximum stock is 11,000 gallons, and if 5,000 gallons at a time, it is 6,000 gallons, the average stock being 6,000 ($1{,}000+\frac{1}{2}10{,}000$) gallons and 3,500 gallons respectively.

It looks as though the average stock for the interest charge, and the maximum stock for the insurance charge provide two more examples of question-begging factors. Neither can be assessed until the size of order or delivery is known. There is no doubt that the formula with the factor definitions as handed down through the literature is impracticable.

Is there an alternative approach to a formula for the EOQ? The answer, so far as I can see, is no. But there is a less general and more direct costing approach. The EOQ is that size of order or delivery that minimises the annual variable cost to the stockholder. For this purpose, the annual cost of holding a buffer stock can be regarded as a constant, i.e. a part of the fixed or time cost, the variable cost being that part of the total annual cost that varies with the size of the order.

The costing approach to EOQ

Costing the road haulage contractor's stock as detailed above is a simple exercise. First there are the three different prices of 20p, 19p and 18p to account for. These give for the total annual price paid excluding the fixed charge, £6,000 (30,000 gallons at 20p), £5,700, and £5,400. The annual cost of the fixed charge depends upon the number of deliveries a year required to maintain the stock at a minimum of 1,000 gallons. If the size of the delivery ordered is 1,000 gallons, then to meet the total consumption of 30,000 gallons a year without depleting the stock in hand, the number of deliveries is 30 a year, for which the annual cost would be 30 at £2 equals £60. For a delivery size of 1,000 gallons, the maximum stock is 2,000 gallons, i.e. a minimum stock of 1,000 gallons plus a delivery of 1,000 gallons. If the insurance rate is £1 per cent, the annual premium on 2,000 gallons valued at the appropriate price of 19p a gallon would be £380 at £1 per cent equals £3.80. Using the same criterion for valuing the average stock to determine the interest charge, we have:

$$1{,}000 + \tfrac{1}{2}\,1{,}000 = 1{,}500 \text{ gallons at } 19\text{p} = £285.$$

If the appropriate rate of interest is 10 per cent, then the annual interest charge would be £28·5. Summarising, we have:

	£
Total price paid	5,700·0
Fixed charges	60·0
Insurance	3·8
Interest on average stock	28·5
Total annual cost when size of delivery is 1,000 gallons	5,792·3

The annual cost for each of a wide range of delivery sizes throws light on the problem at once. The costs can quickly be determined by a systematic approach. Table 17.1 contains a summary. It will be seen that the factor with the greatest force in this example is the price variation. In consequence, the EOQ is to be found amongst those delivery sizes which command the maximum rate of discount.

However, amongst the smaller size deliveries, the fixed charge per delivery is important, and accounts for an annual cost of £600 for deliveries of 100 gallons each. The overall total annual cost given in the last column records a decline as the size of delivery increases, and reaches a minimum for deliveries of 5,000 gallons, whence it shows a steady upward trend. The figure of 5,000 gallons is the EOQ for the haulage contractor's diesel oil. It is the smallest size of delivery at the lowest price charged.

TABLE 17.1 Comparative annual costs for different sizes of delivery of diesel fuel oil

Size of delivery gal.	*Price paid per total gal.* (p)	(£)	*Fixed charge per delivery* (£)	*Insurance of stock* (£)	*Interest on stock* (£)	*Total annual cost* (£)
100	20	6,000·0	600·0	2·2	21·0	6,623·2
800	20	6,000·0	75·0	3·6	28·0	6,106·6
1,000	19	5,700·0	60·0	3·8	28·5	5,792·3
4,000	19	5,700·0	15·0	9·5	57·0	5,781·5
5,000	18	5,400·0	12·0	10·8	63·0	5,485·8
6,000	18	5,400·0	10·0	12·6	72·0	5,494·6
7,000	18	5,400·0	8·6	14·4	81·0	5,504·0
8,000	18	5,400·0	7·5	16·2	90·0	5,513·7
9,000	18	5,400·0	6·7	18·0	99·0	5,523·7
10,000	18	5,400·0	6·0	19·8	108·0	5,533·8

The model

With the annual cost measured on the vertical axis and the size of delivery on the horizontal axis, the cost curve falls to the right until it reaches its trough at a delivery size of 5,000 gallons, where it changes trend to move upward to the right. But the curve is discontinuous.

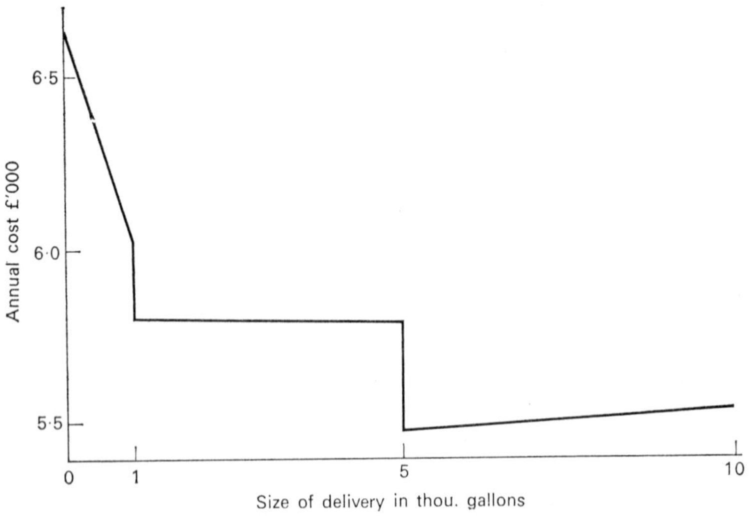

FIG. 17.1 Relation of annual cost to size of delivery

There are two breaks in it, and it consists of three sections, one for each price. Although the example is hypothetical, the curve, which is depicted in Fig. 17.1, is probably fairly typical for materials on which suppliers raise a fixed charge for delivery. Like the curve as a whole, each section has a tendency to be concave upward; but one can easily test the sections for linearity by applying the method of finite differences to the data. Consider the section covering deliveries of 5,000 gallons each and more. Table 17.2 contains the test applied to this upper reach of the curve. As the differences in the size series are all the same and equal to 1,000, there is no need to divide them into the differences in the annual cost series for purposes of the test. These differences, which are given in the penultimate column, are sufficient: they show an upward trend, and the second differences given in the last column, show a downward trend.

TABLE 17.2 Testing for linearity

Size of delivery gal.	*Differences* gal.	*Annual cost* (£)	*First differences* (£)	*Second differences* (£)
5,000	—	5,485·8	—	—
6,000	1,000	5,494·6	8·8	—
7,000	1,000	5,504·0	9·4	0·6
8,000	1,000	5,513·7	9·7	0·3
9,000	1,000	5,523·7	10·0	0·3
10,000	1,000	5,533·8	10·2	0·2

The first differences suggest that the trend of total annual costs in the third column may be a geometric progression, in which case the appropriate form of model would be that of equation (i) of Chapter 7. Table 17.3 contains the test for this form of equation, and it proves

TABLE 17.3 Testing for the exponential form

Size of delivery gal.	*Annual cost* (£)	*log of annual cost*	
		Of total	*First difference*
5,000	5,485·8	3·7393	—
6,000	5,494·6	3·7400	0·0007
7,000	5,504·0	3·7407	0·0007
8,000	5,513·7	3·7415	0·0008
9,000	5,523·7	3·7422	0·0007
10,000	5,533·8	3·7430	0·0008

positive; the common difference is about 0·0007 per 1,000 gallons of delivery size, and the common ratio of the natural numbers is about 1·0015. However, the second differences in Table 17.2 are so small, that it could be justifiably argued that the relationship between annual cost and size of delivery is linear to all intents and purposes. True enough, the annual cost calculated for any intermediate size of delivery such as 5,500 gallons would be much the same whether calculated by the linear equation or by the logarithmic.

Whichever is used, the minimum annual cost is shown to be that where the haulage contractor's standing order is for a delivery of 5,000 gallons, six times a year.

CONCLUSION

It needs to be emphasised that the example of the haulage contractor's standing order for the delivery of diesel fuel oil is entirely imaginary. Nevertheless, there is reason for supposing that where there is a fixed charge per delivery and a discount for bulk deliveries, the general pattern revealed by the analysis outlined above would be fairly common in practice with all kinds of durable materials and goods purchased for consumption, in providing services and in manufacture, or for resale.

There is a limit to the kind of stocks that could usefully be analysed in this way. Analysis is expensive, and the expense must be set against the potential saving in annual costs. It would scarcely pay a small tailor to employ a statistician or a consultant to tell him how much basting cotton he should order at a time and what his minimum stock should be.

Two advantages that the costing approach to EOQ have over a mathematical formula are:

(*a*) It does not need a mathematical proof; and
(*b*) It shows the annual costs for a range of sizes of order quantity, and indicates the effect of a given change in size of order on the annual cost.

Cost minimisation must always be the prime consideration, with convenience to routine a secondary consideration. Where differences in annual costs are insignificant, the convenience of delivery times and frequency can be taken into account as a criterion of EOQ.

So far as (*a*) is concerned, it goes without saying that a management statistician cannot afford to use formulae of which he has no proof. A mathematical proof throws light on the definitions of the independent variables; moreover without a proof, a formula, no matter how good its source and authority, is best regarded as dogma.

On (*b*), it is worth drawing attention to the position in the scale of annual costs against size of delivery of the EOQ. It happens to lie at the smallest size of delivery that commands the lowest price. That is the kind of position where one would expect to find the EOQ.

18

Input–output

Every management statistician operates his own input–output system: the input consists of raw or basic statistics; and the output, of derived statistics such as ratios, averages and parameters (marginal costs, elasticities of demand, optimum prices, economic order quantities and the rest). For some purposes a part of the input is the time spent by the statistician and his staff on operating the system. This is one example of many: it holds true, *mutatis mutandis*, for financial and management accountants, and indeed for every employee from the policy maker to the floor sweeper. For the policy maker, the input consists of the relevant facts; and the output, of policy decisions. For the floor sweeper, the energy he uses or *puts into* his work provides the input, and a clean shop floor provides the output.

Input–output analysis at the individual level is more generally referred to as activity analysis. When the input and output of groups of people working as a unit are being analysed, the work is usually referred to as systems analysis.

INPUT

Input for systems analysis may consist of materials, partly finished goods, data or labour, or a combination of any two or more. Its measure is expressed in terms of money in unit time. Money value is the only homogeneous form in which goods, labour and data can be brought to terms such that they can be added together; and it is necessarily expressed as a rate of flow.

The evaluation of materials is usually a simple matter, provided the supplier is independent of the company. Labour, too, is relatively easy to value. Data, partly finished goods including parts of the company's own manufacture, and materials acquired from an internal source are not so easy to value.

Data may be made up of routine statistics, the compilation cost of which may be entirely non-variable. Even statistics specially compiled for an input–output exercise may not involve the company in additional annual costs. Nine times out of ten they would not.

In a fully integrated system, the transfer price, which should be equal to the marginal cost of production at the supplying department, is the correct price to use for evaluating the input of goods derived from an internal source.[1] Where a fair market price is charged for internal transactions, it is suggested that for evaluating the input of goods from internal sources, the marginal cost or the best approximation to it should be used.

OUTPUT

Similar considerations apply to output. There is one important possible difference, and that is, it may often be correct to value output for internal transfer at a fair market price rather than a marginal transfer price. What the systems analyst is seeking is the optimum system, i.e., the system that will maximise the profits of the company as a whole, which does not always tally with the maximisation of the profits derived from systems or profit centres.

Where the output consists entirely of data, as it does in the offices of the statistician and the accountant, evaluation is almost impossible. Cost-plus is too arbitrary for the purposes of a systems analysis. A fair market price is out of the question. If the company is of any importance in its industry competitors would be willing to pay a very high price for confidential data of the kind produced by management statisticians and accountants as would the company itself for similar information relating to its competitors. It may be thought that if a competitor can put a price on the company's confidential data, the company ought to be able to do it also.

However, market price and utility are two different things. In the case of confidential data, which by definition are processed for internal information, the analyst is concerned with utility and not at all with the price the data would fetch if they were offered for sale. It is the value to the company that provides the criterion, not the potential value to other people. So the question the systems analyst puts to the statistician or the accountant is, 'Where do you send the processed data?' and having

[1] The problem of transfer pricing is discussed in my *Planning Profit Strategies*, Longman, 1971 (Chapter 2).

obtained an answer, he then proceeds to question the recipients: 'What do you use the data for?' The answer may be for policy making, for executive decision making, or for management and control of the business. Whatever it may be, the analyst may have to go further in order to satisfy himself that the output of the system is worth the input including the processing. Sketchy answers like 'policy making' are generally not good enough. What the analyst wants to know is how such data can assist the policy maker.

SYSTEMS ANALYSIS

Systems analysis came into existence as a result of the computer. Before the work of a system could be transferred to a computer, it had to be reduced to a routine. This reduction process was originally the function of systems analysis; but it was so effective in increasing efficiency that it became a management technique in its own right. It has been contended that half the annual savings attributed to computers are really due to the systems analyses. The technique is now employed in sections and departments where the output consists of material things and services; but the same name, systems analysis, is often used; though some firms employ the technique under another name or no name at all.

One of the most difficult items of output to evaluate is general information distributed in a periodical broadsheet or bulletin to executives and managers at all levels. It seems right and proper that such people should be kept informed of what is happening inside the organisation. The question is, What is *right and proper*? Does a wide distribution of production, operating, sales and financial statistics make any contribution to the company's profits? Many managers glance at the figures that most affect them—figures which they themselves may have provided—and ignore the rest. There is always a danger that a copy of the bulletin will fall into the hands of the company's most aggressive and dangerous competitor.

Whether the systems analyst should look into the question of security depends upon his terms of reference. His chief concern is with the profitability of the system. But the security aspect does provide an argument against the compilation and internal distribution of a statistical bulletin. An alternative approach is the preparation and distribution of a statement of the statistical information available to managers, showing frequency of collation, and where the data may be obtained.

A revised statement would probably be necessary from time to time. Regular statistical statements, some exhaustive, some concerned with a sector only, would undoubtedly be needed, so that it is questionable whether the suggested alternative would result in an annual saving.

Once the systems analyst has satisfied himself that the output of the system under investigation is worth while, he can turn his attention to the input and internal operation of the system. Is there a cheaper way of acquiring the input? Is the whole of the input necessary to provide the required output? Is the internal processing the most efficient that can be devised? These and similar questions arise, and it is a function of the analyst to find the answers, and to make suggestions for improvement.

Unfortunately, the literature is much jargon-ridden, some of it, one may have reason for supposing designed to give an esoteric air to the self-evident. There are, for example, the *demand pattern*, the *detector*, *feedback* and *the corrector*, all of which are superfluous when they are understood. The demand pattern is simply the demand for the output of the system and the form it takes. The detector is a purely theoretical box in a diagram depicting the system, the object of the box being supposedly to pick up changes in the output caused by changes in the demand pattern, and to feed these changes back to the corrector, another theoretical box, which is supposed to adjust the input to meet the required changes in output. Could anything be more self-evident? More superfluous? It provides one more example of the academic picture on the box. All systems are dynamic in practice, and therefore all are adjustable to change. Detectors, feedback and correctors operate almost automatically; they go without saying.

Input has a dychotomy of classes, one being described as *main* or *primary*, and the other as *subsidiary* or *secondary*. The distinction between the two is not at all clear. Secondary inputs are variously described as inputs designed to meet changes in environmental or extraneous factors. There is another form of input described as *environment* or *side conditions*, which are said to be more stable than those environmental factors that necessitate secondary input. A large part of the literature on systems analysis calls to mind a dialectic but telling interjection which used to be current in a part of the Midlands when somebody was going into a great deal of self-evident detail; it is: 'Ar says, "Didst?" ' 'Ae says, "Aar" ', of which a translation is 'I said, "Did you?" ' He said, "Yes" '.

None of this is to say that systems analysis is not a good management technique. Far from it. It is the way the subject is presented in the literature that is open to criticism. It is a great pity that it should be wrapped up in so much esoteric jargon.

A BROAD CONCEPT

Input–output has a much broader connotation than that relating to company systems. It is the input–output of each industry of the economy, the input and output being given in detail as to source and destination. Both source and destination consist in part of other industries or industry groups based on the Standard Industrial Classification; sources also include imports and sales by final buyers, and destinations include exports and other final buyers analysed over personal sector, public sector, fixed capital formation and stocks. The input includes labour and such services as transport and communication and those rendered by the distributive industries.

Of all the environmental data a market research worker could wish for, an input–output matrix (or table) for the economy presents the broadest picture. The concept was originally propounded by Professor Wassily Leontiev in the nineteen-thirties for use by the Soviet Government in formulating a national investment programme. Its value for such a purpose will be appreciated; but a matrix for a capitalist economy can be almost as useful to the government, and individual industries and undertakings as the Russian one is to the socialist government for which it was compiled.

Each input–output matrix usually covers a period of a year. Its size, that is, the number of statistical cells it contains, depends upon the number of industries or industrial groups into which the economy is divided for the purpose. It now seems to be the official practice in the United Kingdom to publish a *summary* matrix for each detailed census of production year distinguishing about 30 industries or groups, and a final matrix a year or two later distinguishing upwards of 60 industries or groups. The latest matrices available at the time of writing relate to 1968. The published matrices are described as provisional, the main tables being 'also available in computer readable form'. They distinguish 70 separate industries or groups. Summary tables for 1968, which distinguish 35 industries or groups, were published in *Economic Trends* for January 1971. An explanatory article accompanied the summary tables; but a complete description of the input-output matrices will be found in *Studies in Official Statistics No. 16* (HMSO).

Unfortunately, the collation of input–output tables is an immense task, and only a central government could undertake it. They are compiled from data supplied by individual undertakings in the returns of the detailed censuses of production.

Input–output in the United Kingdom

The first published table for the United Kingdom relates to the year 1948, and not surprisingly it is of academic origin. It is called 'Input–Output Table for the United Kingdom' by L. G. Stewart of the Department of Applied Economics, Cambridge; it appeared in *The London and Cambridge Economic Bulletin*, published in *The Times Review of Industry*, December 1958. Needless to say, the study would have been an impossible task without the assistance and co-operation of the Board of Trade's Census of Production Office, and financial aid through a grant from Conditional Aid Funds.

Perhaps it was all to the good that the trained and not unpractical minds of the academics at the Department of Applied Economics should have been brought to bear on the many major and minor problems that had to be solved in making the table.

Stewart's matrix set the pattern for all official matrices since compiled and published. There are some differences, such as the way imports are treated. Each industry or group has a column and a line (or row) to itself, the former giving the input and the latter the output.

As well as for 1948 there were detailed censuses of production taken for each of the years 1950, 1951, 1954, 1958, 1963 and 1968; and presumably there will be another for 1973. No detailed input data were collected for 1951, but input–output matrices have been compiled for the other years up to 1968. For 1954, the final tables were published in 1961, in *Studies in Official Statistics, No. 8*; but since then the time lag has been reduced from seven years to about three.

Practical applications

The value of an input–output matrix to the individual company depends upon the size of the company and its organisation. A glance at the matrix for any year suggests a system of business ratios consisting of the ratio of sales to each individual industry or group to total sales. A comparison of the company's ratios with the ratios derived from the matrix for the same year might bring to light some untapped markets for the company's products. Care needs to be taken with definitions: the notes accompanying published matrices should be examined carefully. For instance the tables exclude intra-industry transactions, so that the sales of any industry to itself are shown as zero. This is one problem of definition especially to those companies which sell a large proportion of their output to other companies in the same industry. For the purpose of a system of business ratios, a simple adjustment to the company's total sales to provide the appropriate denominator would probably be enough to solve the problem.

A similar system for the input side of the business could as easily be compiled, but as a rule, it would scarcely be worth the trouble. However, the input statistics for the industry, i.e., the figures down the column, should be of interest and value to the company's stores department: they might bring to light the way the industry was tending to exploit new materials and new sources of materials. They might even suggest a more scientific approach to the buying function of the stores department.

The Leontiev concept applied within the company

Where a company consists of a number of divisions, i.e. departments or companies of a group company, it is possible to draw up a central control

TABLE 18.1 Form of central control chart £'000

						External		
Purchases by . . .	*A*	*B*	*C*	*D*	*Total*	*Home sales*	*Exports*	*Total sales*
Sales by:								
A	. .							
B		. .						
C			. .					
D				. .				
TOTAL					+			
External								
Home						—	—	—
Imports						—	—	—
Total purchases						—	—	—

\+ Sum of both down and across.
. . = zero.
— = not applicable.

chart on the lines of Leontiev's inter-industry formula. An inter-division control chart is discussed in Chapter 2 of my *Planning Profit Strategies*. Briefly, such a control chart could usefully include columns for external sales subdivided between home and export and lines for external purchases subdivided between home purchases and imports. The matrix would appear something like Table 18.1 in which A, B, C, D represent divisions of the company or group.

Whether the chart should be compiled monthly, quarterly or annually is partly a matter of convenience, which depends upon the frequency with which routine statistics are compiled. Central managements alive to the necessity of integration within the company are likely to call for the chart more frequently than those which are not.

It may be thought at first glance that the figures in the upper right-hand sector of the internal part of the table would be the same as those in the lower left-hand sector. But this is not necessarily so. The cell of line A column B would give the sales of A to B and therefore the purchases of B from A; where as the cell for line B column A would give the sales of B to A, and therefore the purchases of A from B.

A period-by-period comparison indicates the extent to which inter-divisional trading is expanding or contracting relatively to changes in external transactions. This would give central management an idea of the way the degree of integration was changing. No company can ever hope to be self-contained in its demand for materials and services, there must always be some external purchases, and the proportion that the value of these bear to the total purchases must depend in some measure upon relative prices.

Another problem that may arise is the valuation of internal transfers. Should it be based on the actual transfer price, or upon some notional fair market price? Where a rational transfer pricing system is in operation, the transfer price, which is equal to the marginal cost, would be much lower than any notional fair market price. Perhaps the choice of criterion is not important, so long as once made, it persists.

If the external sales of one of the divisions of the company are expected to change, the internal sales of other divisions will be affected, and this will result in further changes in internal transactions. With the aid of an input–output control chart, it would be possible to trace these changes to their ultimate, and what is more, to measure them, and so enable management to forecast the demand on internal and external sources for materials and intermediate products.

19

Break-even analysis

A static form of break-even analysis as propounded by linear programming was briefly discussed in Chapter 5. It is now proposed to go into the problem more deeply. Some of the equations used have been given in earlier chapters, notably Chapters 5, 8 and 11; but they will be restated here for the convenience of the reader. It will be necessary to refer back to two of the diagrams given in Chapter 5; two or three more will be given in this chapter. As implied in Chapter 5, there are two distinct approaches to break-even analysis, or profit planning as it is now often called. One is the linear and the other the curvilinear.

THE LINEAR APPROACH

Linear break-even analysis is demonstrated in Fig. 5.1. It shows the graphs of the annual cost equation marked C and the linear annual gross revenue equation marked S, which assumes that changes in the rate of gross revenue results from changes in the quantity sold under conditions of a fixed unalterable price.

As we have seen, the total cost equation is written:

$$T=aQ+F \qquad \ldots\text{(i)}$$

where T is the total annual cost; a, the marginal cost; Q the quantity sold in a year; and F, the fixed or time cost. (It is worth noting that in notation used throughout this chapter, capital letters denote annual rates or totals in unit time; and small letters, ratios, averages and the like.) The linear sales-proceeds equation is:

$$S=pQ \qquad \ldots\text{(ii)}$$

in which S represents the sales-proceeds; and p, the price. It can be said that the product breaks even when $S=T$; that is, when:

$$aQ+F=pQ \qquad \text{. . . (iii)}$$

From this, we have:

$$F=Q(p-a) \qquad \text{. . . (iv)}$$

In the jargon, $p-a$ is called the unit contribution, so that $Q(p-a)$ is the total contribution, written C. In addition, aQ is called the variable cost, and written V. With P for profit, we then have $P=0$ where $F=C$. In any event, $P=S-T=S-(V+F)=C-F$; and $C=S-V=F+P$. The annual (or unit-time) rates represented by the capital letters shown appear to provide the standard notation of the linear analysis of the business schools.

We can take in turn each factor in equation (iii) as the variable leaving the remainder fixed. Using subscript v as denoting the variable, we have from equation (iii) or (iv) the following:

$$F_v=S-aQ \qquad \text{. . . (v)}$$

$$p_v=\frac{aQ+F}{Q} \qquad \text{. . . (vi)}$$

$$a_v=\frac{pQ-F}{Q} \qquad \text{. . . (vii)}$$

$$Q_v=\frac{F}{p-a} \qquad \text{. . . (viii)}$$

Of these equations, only (viii) tells us something new, the remainder being self-evident truths. Equation (viii) means that the break-even rate of output, Q_v, is equal to the fixed cost divided by the unit contribution, $p-a$.

It follows that with Q as the variable factor:

$$S_v=pQ_v=\frac{pF}{p-a}$$

Multiplying both the numerator and denominator by Q, we have:

$$S_v=\frac{SF}{Q(p-a)}=F\frac{S}{C} \qquad \text{. . . (ix)}$$

which in linear analysis is sometimes written:

$$S_v=\frac{FS}{S-V} \text{ or } \frac{FS}{C} \text{ or } F\frac{S}{C}$$

Equation (ix) means that $F/C=1$, which, since in a break-even situation $F=C$, is perfectly true. However, our concern will be with the ratio S/C, which will be examined later.

THE CURVILINEAR APPROACH[1]

We can accept the linear total cost curve of equation (i) as written. We need to substitute for the linear sales proceeds, equation (ii), an equation which will take account of the effect on S of a changing price. Taking the elasticity of demand to be a constant within the observed range of prices, we have:

$$Q=k\,p^{-e} \qquad \ldots \text{(x)}$$

where k is a constant and e is the price-elasticity of demand. It is worth mentioning that this forms the basic model, other factors being added as necessary, used in the regression analysis of market demand.

Then:

$$p=K\,Q^{-1/e} \qquad \ldots \text{(xi)}$$

where $K=k^{1/e}$. Multiplying both sides by Q, we have:

$$S=pQ=K\,Q^{(1-1/e)} \qquad \ldots \text{(xii)}$$

Curvilinear analysis sees the problem in terms of maximising profits, rather than of seeking a break-even price. As we have seen, the optimum or rational price is:

$$p_m=\frac{ae}{e-1} \qquad \ldots \text{(xiii)}$$

where subscript m denotes the optimum.

However, there is a break-even situation demonstrated in Fig. 19.1 with optimum price where the cost curve is tangential to the sales curve, for which the quantity sold must be regarded as the variable factor, and may be written Q_v. We then have p_m = the unit cost, that is:

$$\frac{ae}{e-1}=a+F/Q$$

so that

$$F=aQ\left(\frac{ae}{e-1}-1\right)$$

and

$$Q_v=\frac{F(e-1)}{a} \qquad \ldots \text{(xiv)}$$

[1] I first developed the curvilinear approach in a paper on 'The Role of the Price-elasticity of Demand in Profit Planning'. *The Journal of Business Finance*, 1971.

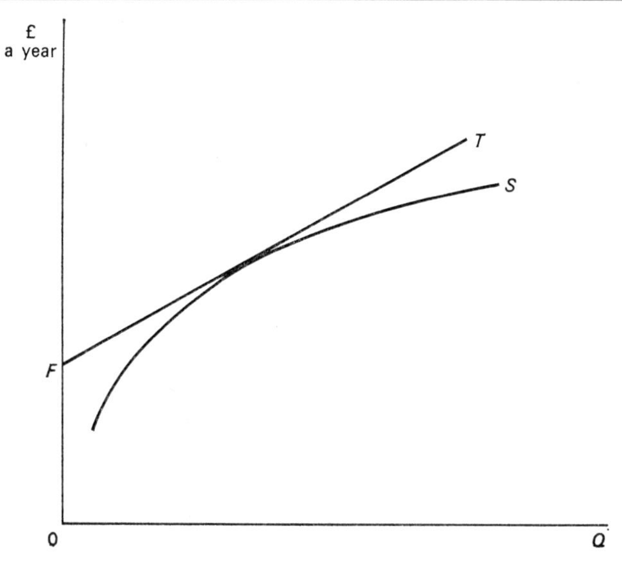

FIG. 19.1 A break-even situation with optimum price

It follows that:

$$S_v = p_m Q = \frac{ae}{e-1} \cdot \frac{F(e-1)}{a} \qquad \ldots \text{(xv)}$$

$$= Fe$$

Since both F and e are entirely independent of Q and p, equation (xv) holds true whether Q and p are at their optima or not. Equally, it applies to $p = p_m$, and gives the break-even sales-proceeds for $Q = Q_v$. For the marketing policy purposes, equation (xiv) is probably more useful than (xv), though admittedly, an increase in marketing activity designed to raise Q and S, may affect F and a. If F and a change in the same proportion as each other, the calculated value of Q_v would remain unchanged.

THE TWO BREAK-EVEN PRICES

It will be seen from Fig. 5.4 that when there are positive profits, there are two break-even situations, marked B_1 and B_2. There are then two break-even prices, one above the optimum and the other below. It is not possible to imagine any circumstances in which an organisation might wish to reduce its profits by charging prices above the optimum;

but such non-profit seeking organisations as the state industries may be interested in the lower break-even price.

A break-even price formula in this respect appears to be unobtainable; so that the only means of ascertaining either or both of the two break-even prices is to approach it by means of a net revenue schedule. A hypothetical example can be used for demonstrating how this is done. A company makes and sells a product brand whose marginal cost is £3 and whose price-elasticity of demand is two. The optimum price is:

$$\frac{3\times 2}{2-1} = £6$$

Working down from this to the lower break-even price, we have the profit schedule shown in Table 19.1. The value of the constant, k, in the equation expressing the annual quantity sold as a function of price—equation (x) above—is 63,500, and the annual fixed cost represented by F in equation (iii) above, is £4,000.

The table shows that £4·00 is rather lower than the break-even price, and that £4·10 is rather higher. Working down from £4·10 we finally arrive at the break-even price expressed to the nearest penny, of £4·02.

TABLE 19.1 Approach to the lower break-even price

	Annual sales			
Price p £	*Quantity* Q No.	*Receipts* $S = pQ$ £	*Annual cost* $T = 3Q + 4{,}000$ £	*Net revenue* $S - T$ £
7·00	1,296	9,072	7,888	1,184
6·00	1,764	10,584	9,292	1,292
5·00	2,540	12,700	11,620	1,080
4·00	3,969	15,876	15,907	−31
4·10	3,778	15,482	15,328	154
4·08	3,815	15,565	15,445	120
4·06	3,852	15,639	15,556	83
4·04	3,891	15,720	15,673	47
4·02	3,929	15,795	15,787	8

THE IMPLIED PRICE-ELASTICITY OF DEMAND

It can be said that for any price of a product there is an implied elasticity of demand, the assumption being that the actual price is the optimum. As I point out in my *Appraising Capital Works*, the 'implied elasticity is equal to the actual price divided by the actual price minus the marginal cost', that is:

$$e_i = \frac{p}{p-a} \qquad \ldots \text{(xvi)}$$

where e_i is implied demand elasticity. Multiplying the numerator and the denominator by Q, we have:

$$e_i = \frac{pQ}{Q(p-a)}$$

$$= \frac{S}{C}$$

(Equation (xvi) is derived from the rational price formula of equation (xiii).)

Thus, the linear and curvilinear approaches converge to a common break-even formula. We can rewrite equation (ix):

$$S_v = Fe_i \qquad \ldots \text{(xvii)}$$

so that where the fixed price used in the linear analysis is rational, then $e_i = e$, and equations (ix) and (xv) give the same solution values of S_v. It should be added that linear programming literature does not make any but passing reference to the elasticity of demand or to the rational or optimum price, it being assumed, presumably, that the firm's pricing-policy is independent of profit planning.

WHAT IS FIXED COST?

If break-even analysis is as important as linear-programming literature implies, and if we are to determine break-even sales proceeds by calculating FS/C or Fe, then F, the annual fixed cost, must be clearly defined. With a single-product company, there is no difficulty: F is equal to the total annual cost normally incurred by the company, including directors' fees, interest on loan capital and bank overdraft, and depreciation, less the annual variable cost, V. It would exclude investment expenditure of all kinds, including that on below-the-line advertising. The total annual cost as defined is entirely attributable to the single product, so that:

$$F = T - V \text{ or } T - aQ$$

where T is the annual cost.

With multi-product companies there are some serious difficulties. First, there is the problem of allocating fixed costs amongst the several products. Management accountants employ a variety of formulae in product costing. The formulae have one thing in common: they are all arbitrary in the allocation of fixed costs. Full-blooded absorption costing has tended to fall into disrepute since marginal costing began to be accepted; but unit-costing is still generally practised. It is now direct: under absorption costing, the top management salaries bill, for instance, was allocated not directly to products, but over the various departments, production, marketing, purchasing, personnel, secretary's, accountant's, transport and distribution, in accordance with some formula such as *pro rata* to total departmental expenditure or payroll. The non-productive departments would each then pass on their total costs including those allocated from above, to the production department, or in some cases, separate from them the materials and labour and pass the former to the stores department and the latter to the personnel department, which in turn would pass them on to the production department's account. The point of this whirligig procedure was, it seems, to build up total departmental costs including allocations from above, and to make sure that every single item of production bore its appropriate share of the company's fixed costs.

Direct labour and materials were always borne by the several products; but the introduction of marginal costing, which embraces rather more than direct labour and materials witnessed a change. Now, the modern product-costing technique is the more sensible and certainly the less expensive one of allocating all fixed costs direct and *pro rata* to the variable cost: that is:

$$F_t = T - \sum V$$

where F_t is the company's total fixed cost and $\sum$ is a summation sign. Then the fixed cost of product x may be written:

$$F_x = \frac{V_x(T - \sum V)}{\sum V}$$

or, for the unit fixed cost:

$$F_x/Q_x = \frac{a_x(T - \sum V)}{\sum V}$$

In practice, of course, the calculated fixed cost is expressed as a percentage addition to the variable cost or marginal cost to give the total cost or unit cost.

Attribution costing

Although modern practice is more sensible and less expensive than absorption costing, it is none the less arbitrary. Is there a more disciplined procedure? Attribution costing goes some way towards providing an answer. An attributable cost is a cost that can be directly attributed to the particular product. The variable and marginal costs are attributable, so too are certain fixed costs, for the most part those related to plant capacity. With technological progress, general purpose machinery and plant are giving way to the more efficient specialist types of fixed assets, so that attributable fixed costs, consisting for the most part of annual capital charges, and repair and maintenance costs are becoming a more significant part of total fixed costs. It would be true to say, however, that some general purpose items of capacity may be limited in use to a particular product, and that some specialist items may be used in the manufacture of more than a single brand of product. However, these restraints on an acceptable connoted definition of attributable fixed costs do not prevent us from accepting a conceptual denoted definition of fixed costs attributable to *specialist* units of capacity, that is, specialist in the sense of limitation in use to the single brand, and therefore attributable to the brand.

For the single brand, we may divide the fixed costs into the two elements:

$$F = bn + F_h \qquad \ldots \text{(xviii)}$$

where b represents the annual cost of a unit of specialist capacity; n, the number of units of specialist capacity available for use; and subscript h denotes the hard core of non-attributable costs arbitrarily allocated, in accordance with the company's standard formula, to the product. It is convenient to assume that the several units of capacity are each of the same annual cost and production capacity. If the normal annual capacity of each unit of capacity is N units of output, then the horizontal axis of our annual sales-proceeds and total cost diagram can be calibrated for the number of units of capacity: $N, 2N, 3N \ldots nN$.

There is a school of thought which speaks of the short-term and long-term marginal costs, the former being a, as defined above, and the latter, a concept that takes account of changes in fixed costs resulting from changes in capacity. The latter, it seems, provides for a step in the annual cost curve to accommodate each additional unit of capacity.[1]

[1] The Institute of Cost and Works Accountants in its *Report on Marginal Costing* (1961) falls into this trap; *vide* pp. 46–7.

It is difficult to reconcile this idea with the realities of business practice. If the actual annual output or quantity sold calls for the use of n units of capacity, and if Q falls to a rate which necessitates the use of only $n-1$ units of capacity, such a stepped cost curve would imply that the annual fixed costs of one unit of capacity would be saved. Except where it is believed that the chances of restoring the rate of sales are insignificant, such an idea is quite indefensible: no company deliberately reduces its plant capacity without good reason. Once n units of capacity have been installed, they are likely to remain; additions may be made but not displacements. In short, with an additional unit of capacity, the total cost curve rises to a new position throughout its length. With each successive addition, it rises, and lengthens to a new normal capacity.

That is consistent with a total annual cost equation embracing the fixed-cost formula of equation (xviii):

$$T = aQ + bn + F_h \qquad \ldots \text{(xix)}$$

It is represented in a three-dimensional total cost diagram, in which the third dimension consists of a cost curve that rises bodily throughout its length with every increase in capacity. Provided the marginal cost of the product remains unchanged, the several cost curves will be parallel to each other. The hard-core fixed cost, F_h, is represented by

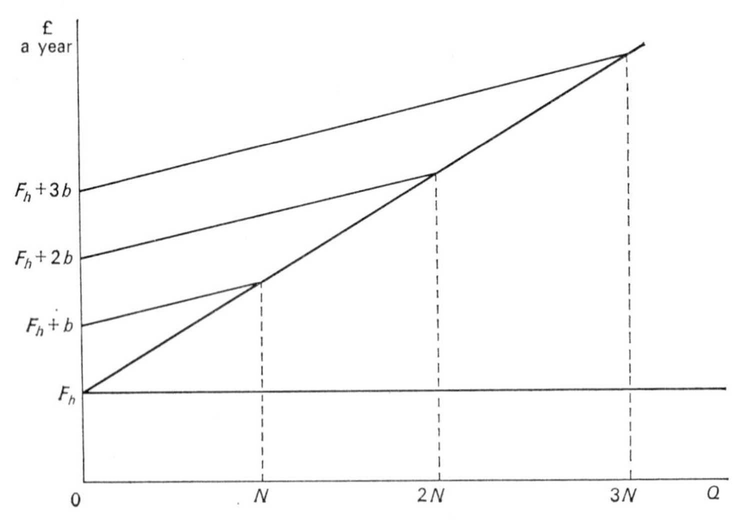

FIG. 19.2 The cost curve and capacity change

a horizontal straight line beginning on the vertical axis at F_h. The total cost curve for $n=1$ will rise from a point on the vertical axis equal to F_h+b; for $n=2$, F_h+2b, and so on. Just as a is the marginal cost of the product, so b is the marginal cost of the capacity. The curve for $n=1$ will terminate on the right at a point equivalent to N measured on the horizontal axis; that for $n=2$, at a point equivalent to $2N$; and so on, as in Fig. 19.2. On the assumptions made, the terminal points would fall on a straight line which disects the vertical axis at F_h. If there can be such a thing as a long-term marginal cost curve, that is it. But what does it mean? How can it be defined in terms of marginal analysis? Its slope is the marginal cost neither of the product nor of the capacity. It has meaning between the terminal points only if the capacity could be changed by infinitely small amounts.

We can now make an attempt to answer the question posed at the head of the previous section: What is fixed cost? For any purpose, the attributable fixed cost would count. But let us consider the three possible choices open to a profit planner.

First, there is the one in which all fixed costs are completely abandoned. Then for break-even analysis, F would count as zero, and $S_v=Fe_i=Fe=0$. The cost curve would pass through the origin. The break-even quantity sold would also be zero. In linear analysis, the sales and cost curves would intersect at the origin, which would satisfy its method. Curvilinear analysis, however, stipulates that the break-even point exists where the linear cost curve forms a tangent to the sales curve, so that it should apply to the optimum price. This tangential requirement cannot be satisfied where $F=0$. That F could be zero in fact and not merely assumed away is more than a theoretical possibility. It actually exists amongst men and groups of men who offer themselves as contractors undertaking such skilled craftsmen's work as thatching and, now, as a result of the selective employment tax, skilled and labouring work on building sites. In these cases, standard demand and sales curves could scarcely exist; but they could exist for some primitive craft industries where the only capital assets consist of a few loose tools and possibly some form of shelter for which the annual capital charges and repair costs are so insignificant as to be zero for all practical purposes.

Secondly, the analyst may choose to include attributable fixed costs and to omit non-attributable costs. Many, probably most, industrial investment appraisers would approve of this, especially if they have been brought up in the school of thought that regards all annual revenue or savings in excess of project-variable costs as additional annual net revenue and not merely as 'contribution'. Project-variable and attribution costing are not quite the same concept; but the underlying theories are much the same. Both, like marginal costing, reject unrealistic conventions and practices and attempt to steer clear of the arbitrary. Attribution costing is largely, if not entirely, concerned with changes in production, sales and distribution capacity, whereas project-costing is concerned with capital projects of all kinds including technological saving projects as well as expansion. Of the six broad kinds of fixed investment categorised in my paper on 'An Economic Analysis of Fixed Investment',[1] only in one can the fundamental costing principles be described as identical to those of attribution costing, and that is the expansion development category. Expansion reorganisation achieves increased capacity through a more efficient arrangement of existing fixed assets. Saving reorganisation, however, includes, by definition, displacement projects, some of which would involve a reduction in capacity. Here, too, project costing would be much the same as attribution costing.[2] There is much to be said in favour of attribution costing and leaving out of account all non-attributable costs in project planning. It is a dynamic form of marginal costing which recognises that capacity changes are correctly taken care of in an upward and downward movement of the length of the total cost curve.

Finally, the planner's choice may fall to the inclusion of all fixed costs as allocated by the cost accountant. In this event, a division of time costs between attributable and non-attributable would scarcely serve any useful purpose for break-even analysis; but the planner may choose to make a correction to the cost accountant's product costing in the light of such a division. Incidentally, logical and useful as attribution costing is, it is, in common with project-variable costing, almost unknown in British industry. There is no reference to it in *Terminology of Cost Accountancy* compiled by the Institute of Cost and Works Accountants.[3] There are two references in the *Terminology* to marginal costing: one under 'Marginal Costing' and the other under 'Marginal Cost'. Although the Institute now sets itself out to be the authority

[1] *The Economic Journal*, December 1958.

[2] A chapter in my *Appraising Capital Works* is devoted to project costing, which is a recurring subject throughout the work.

[3] I cannot claim to have much knowledge of costing accounting literature. There is only one work I can find that refers to it and that is *Cost Accounting Analysis and Control* (Irwin 1967) by Gordon Shillinglaw, Professor of Accounting, Graduate School of Business, Columbia University.

on management accounting, the *Terminology* makes no reference at all to marginal revenue or optimum or rational pricing.[1] Nor is there any reference to discounted cash flow, although of the three main techniques, the Institute has expressed a preference for the internal rate of return, which is the least practical, and the most illogical, inflexible and unrealistic of the three.[2]

Attributable time costs consist largely of annual capital charges, that is, interest on outlay and depreciation. There is a case for including an interest charge even where there is no loan capital and no bank overdraft, a possible exception being where expansion-development works are financed out of new equity capital, a situation which is somewhat theoretical. Interest on outlay at a rate at least equal to the risk-free rate—the opportunity cost of capital—seems to be reasonable where works are financed mainly out of retained profits. It is a controversial subject which need not concern us here.

Of the three main discounted cash flow techniques, the internal rate of return takes care of depreciation on a sinking fund basis in the discounting process and omits interest on outlay; the net present value takes care of both interest and depreciation in the discounting process, depreciation being included on a sinking fund basis. Only the third technique, annual value, actually calculates the annual capital charge, depreciation on a sinking fund basis, and interest separately if need be, though it can be omitted entirely. Clearly, then, the profit planner has no choice—if he wishes to take account of the attributable time cost separately from the non-attributable time costs, he has no alternative to the annual value technique of cash-flow discounting for calculating the annual capital charges.

PROFIT PLANNING

Whether the profit planner should employ the linear analysis expounded by the schools of business studies or the curvilinear analysis which academic economists would undoubtedly favour, depends largely upon the circumstances. In all such cases as those involving a new product, for which there is no means of ascertaining the elasticity of demand, there is much to be said for the linear approach. The solution value of $S/C = p/(p-a)$, is the implied elasticity of demand, e_i. Anyone

[1] The *Terminology* was published in 1966. A year or two prior to this, the Institute changed the name of its monthly organ from *The Cost Accountant* to *Management Accounting*.

[2] An impartial comparative analysis of the three techniques will be found in Chapters 5 and 6 of *Appraising Capital Works*.

familiar with the elasticities of demand for brands of similar products would be able to say whether the solution value seemed reasonable or not. But the obvious approach to the pricing problem in this kind of situation is to make an estimate of the elasticity of demand with an estimate of the margin of error. Three estimated optimum prices could then be determined, one for each extreme and one for the middle value. With these, the profit planner could then calculate the break-even rate of sales by quantity and value, and so arrive at the annual profit that would be made from the forecast sales, or the quantity that would have to be sold to make the new product or the additional capacity profitable enough to be worth while.

Where the price-elasticity of demand is known, there is not much doubt that curvilinear analysis is superior to the linear. For one thing, Fe provides a unique base, whereas Fe_i does not. Although Fe gives the break-even sales proceeds for the optimum price, unlike Fe_i, it is independent of the actual price. Whereas the linear approach is based on a fixed unalterable price, whether rational or not, the curvilinear approach takes account of changes in sales proceeds resulting from changes in price. It traces the course of the sales-proceeds curve for relatively high prices at the left-hand end to relatively low prices at the right-hand end for a range of prices.

The linear method appears to have one advantage over the curvilinear: changes in quantity that are independent of price change do not affect the position of the linear sales curve. With the curvilinear method, they would cause the sales curve to rise or fall bodily, as in Fig. 19.3. If the optimum price is known and adopted, then so long as the marginal cost and the elasticity of demand remain constant, there is no reason why the simpler linear method should not be adopted, when S/C and $p/(p-a)$ would be equal to the true elasticity of demand. However, it will be seen from equation (ii) that the slope of the linear sales curve is the measure of the price. We could superimpose a straight line from the origin on the curvilinear diagram to represent price. If it were the rational, it would pass through the curvilinear sales curve at a point where the marginal revenue equals the marginal cost. It would indicate at once the new position of the sales curve at the critical point when an independent change in quantity has taken place. The new maximum profit would, as with the old, be measured by the vertical distance between the cost and revenue curves. By this means the two methods are reduced to one, and the essential features of both are retained. Figure 19.3 provides a diagramatic representation of this.

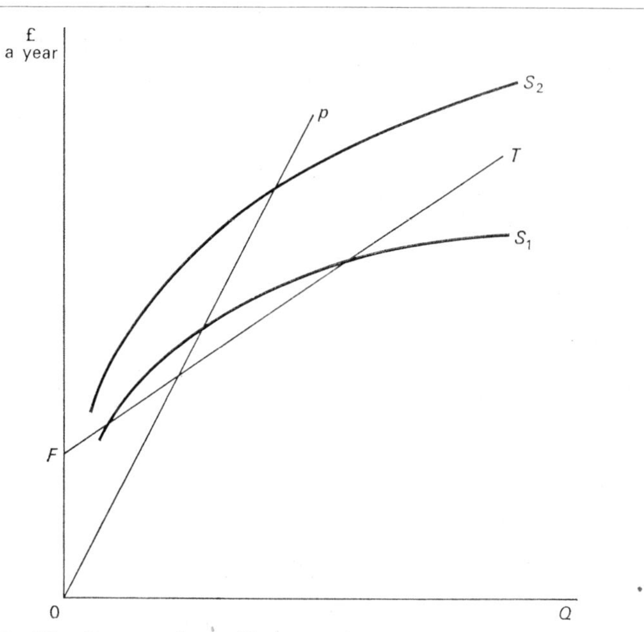

FIG. 19.3 The linear and curvilinear methods and change in the level of sales

As to the relevant definition of fixed cost and whether it should be wholly or partly accounted for in profit planning, the idea of total exclusion can be abandoned on the grounds that the result would be entirely meaningless. Whether the definition of *F* for profit planning purposes should include non-attributable costs or not, there is a school of thought which totally rejects non-attributable costs, or project or capacity non-variables as they may be called, as irrelevant. The principles of cost accounting are not a matter of religious belief but of logical realism. Before the general acceptance of marginal costing, product costing was governed by conventions and practices many of which could scarcely be described as reasoned. Old customs die hard. Costing plays such an important part today in the business life of the community, that cost-accounting research is well worth while. A certain amount has been carried out by the Institute of Cost and Works Accountants but not nearly enough. There is appreciable scope for research into project-variable costing, capacity-variable costing and variable costing in other areas where time costs are involved. After all, much of it would comprise a logical extension of product-marginal costing into the field of time costs. A research team could well consist of members of the Institute's technical staff, academic economists, research workers from the schools of business studies, industrial economists, and commercial accountants.

It is worth mentioning that the linear analyst speaks of what he calls the *P*/*V* ratio,[1] which is defined as 'the profit/volume ratio', that is, in our notation:

$$\frac{100\,(S-V)}{S}=\frac{100C}{S}$$

Why it is called the *P*/*V* ratio is not clear; the *C*/*S* ratio would make more sense within the notation used by the schools of business studies. It is described as 'a measure of the profit that will be earned for an increase in sales of £100'.[2] It will be seen that the ratio is equal to £100 divided by the implied price-elasticity of demand, or the actual elasticity if the price is the optimum.

The difference, $S-S_v$, is called the margin of safety, and the profit on this margin is equal, in our notation, to:

$$\frac{S-S_v}{e_i} \quad \text{or} \quad \frac{S-S_v}{e}$$

as the case may be. Similarly, the profit to be derived where the plant is being used to its full normal capacity is:

$$\frac{pnN-S_v}{e_i}$$

or:

$$\frac{p_m nN-S_v}{e}$$

In a situation like that represented by the former of these two formulae, one way that might be open to the firm for increasing the profit from the product is to raise the price. In either situation, if the actual price is already equal to or greater than the optimum, a price rise could be effective in increasing the profit only if the demand for the product at the actual price exceeds the normal capacity of the plant. Then in practice, the firm would probably increase the rate of output towards the economic capacity of the plant, where the marginal cost is equal

[1] *Vide*, for instance, M. J. Clay and B. H. Walley, *Performance and Profitability*, Longmans, 1965.
[2] *Ibid.*

to the unit cost, which would justify an increase in actual price. But then it would be time to consider installing an additional unit of capacity.

Numerical examples

A few numerical examples will help towards an understanding of linear and curvilinear profit planning. Suppose a product has an elasticity of demand of 3, a marginal cost of £1 and an annual fixed cost of £500, the optimum price would be:

$$\frac{3\times 1}{3-1}=£1{\cdot}5$$

We might then have a break-even situation like that of situation A below:

		Unit	Situation A	Situation B
Q	Quantity of sales in units	No.	1,000	1,333
S	Sales proceeds @ £1·5 a unit	£	1,500	2,000
V	Variable cost @ £1 a unit	£	1,000	1,333
C	Total contribution, $S-V$	£	500	667
F	Fixed cost	£	500	500
P	Profit O − F	£	—	167
$S/C=$e		Ratio	3·0	3·0
$S_v=FS/C=Fe$		£	1,500	1,500

Situation A breaks even, and B shows a profit of £167, which arises from an increase in sales. A and B represent the linear approach with a fixed price which is the optimum.

Suppose now the company finds itself in the break-even situation represented by A. Not being aware that the price of £1·5 is the optimum, it decides to try to achieve a profit by price manipulation and raises it to £2. From equation (x) above, we have:

$$k=Qp^e$$

Substituting, we have:

$$k=1{,}000\times 1{\cdot}5^3$$

$$=3{,}375$$

Then applying equation (x) to $p=2$, we have:

$$Q=\frac{3{,}375}{2^3}$$

422 units

Then with the same level of demand, we have situation C:

		Situation	
		C	*D*
Q	*Units*	422	500
S @ £2 a unit	£	844	1,000
V @ £1 a unit	£	422	500
C S−V	£	422	500
F	£	500	500
P C−F		−78	—
S/C	Ratio	2	2
$F\ S/C$	£	1,000	1,000
Fe	£	1,500	1,500

The company expects a fall in the quantity sold, but believes the fall will be more than offset by the rise in price and the fall in the variable cost. However, expectations prove to be unfounded, since the price now is the sub-optimum one of £2. The company has succeeded in breaking away from the break-even situation, but in the wrong direction, into a loss situation. It decides to retain the price of £2, and to try raising the level of demand by increasing its advertising and marketing activities. Suppose situation D is the result: a new break-even situation with sub-optimum price.

CONCLUSION

A number of conclusions can be drawn from this analysis.

1. The formula of equation (ix), i.e. $S_v = F\ S/C$, is limited to a fixed unalterable price. The value of S/C differs from the elasticity of demand except where that price is the optimum. Only Fe gives a unique solution; the value of $F\ S/C$ depends upon the extent to which the price charged differs from the optimum price. It can be argued that the true value of the break-even sales proceeds is equal to Fe. Where p is greater than p_m, $F\ S/C$ falls short of Fe, and where it is less than p_m, it exceeds Fe.
2. Fe is independent of all factors other than F and e. It remains valid for all prices, all marginal costs, and all quantities of sales.
3. The value of the ratio S/C is a constant where the price and marginal cost remain unchanged. Since $S = pQ$ and $C = Q(p-a)$, it is equivalent to $p/(p-a)$: It is equal to the elasticity of demand only where the price is the optimum, but it can be described as the implied elasticity of demand.

4. Where a company raises a product from a true break-even situation by increasing its marketing activities, the increase in sales proceeds is equal to $S-Fe$, and the profit is equal to this increase less the increase in the variable cost of sales, that is,

$$P=S-Fe-a(Q-Q_v)$$

It may be stated by way of general conclusion that the commonly expounded linear break-even analysis and profit planning demonstrated in equations (i) to (ix) above, seem to oversimplify what amounts to a complex problem, which can be solved only by curvilinear analysis. The proponents of the linear approach should at least make it clear that their break-even point is not unique, and that it can be the true one only by accident. It regards sales proceeds, S, as a linear function of Q, the quantity sold, with p, the price, serving as a coefficient of Q, as equation (ii) and Fig. 5.1 show. But price is very much an independent factor, on which Q itself largely depends. Although the sales proceeds of a brand are always equal to pQ, an increase in p will, other things being equal, almost invariably have the effect of reducing the value of pQ, and *vice versa*, through the effect of a change in price on the quantity sold.

In spite of this, however, the linear analysis is not without practical value. Used in double harness with the curvilinear analysis, it can at least help the management statistician to an understanding of the fundamental principles of break-even analysis and profit planning.

20

Miscellaneous tasks and applications

Oddly enough, the management statistician's work does not always begin and end with numbers: it may begin with making a classification and end with writing a report. It is proposed to discuss these two subjects in turn.

CLASSIFICATION

Classification is fundamental to all learning and knowledge. Without classification there can be neither knowledge nor understanding. So far as is known the ancient Greek philosophers made the first attempt at deliberate classification: they tried to classify everything. But we all classify everything we see, hear, feel, taste or smell instinctively, and we do it from the moment our senses begin to function soon after we are born. Our early classifications are not very scientific, but they are good enough to encourage us to think and to try to understand.

Since the Greek philosophers' categories, the subject has exercised the minds of scientists the world over. A recent biography of Carl Linnaeus, the Swedish scientist (1707–78)[1] reminds us of the great general interest in classification that Linnaeus occasioned when he published his classification of plants. Earlier classifications had been made by reference to the form and arrangement of the petals. Linnaeus classified plants by reference to the structure of the pistil and stamens of the flower, and so to the sexual habits of plants. To him the propagation system of plants was of the first importance, far more important than the flower petals, which, he declared, contribute nothing to generation, but merely serve as the bridal bed. Needless to say, the new classification gave rise to much controversy, and some critics wondered

[1] Wilfred Blunt, *The Complete Naturalist: A Life of Linnaeus*, Collins, 1971.

whether botany could be regarded any longer as a suitable subject for schoolgirls.

The author of the biography, Wilfred Blunt, seems to believe that Linnaeus is now almost forgotten outside his native Sweden, but this is probably a mistaken belief. To most of those who have given the subject any thought, classification is Linnaeus, and Linnaeus is classification: one cannot think of one without thinking of the other.

Management applications

Before turning to consider the attributes of a good classification, and how to devise one, let us take a look at the subject in its practical applications in the field of management and management statistics.

Although business classifications are not made with the object of giving the compiler or the user a greater understanding of the genus that is divided into classes, they do, in fact, do precisely that. Over a hundred years ago, a well-known logician of his time wrote:

> The first efforts in the pursuit of knowledge, then, must be directed to the business of Classification. Perhaps it will be found in the sequel, that Classification is not only the beginning, but the culmination and the end, of human knowledge.[1]

Even so, business classifications have a much more mundane purpose than the pursuit of knowledge. Profit maximisation is the ultimate purpose; the acquisition of knowledge is incidental, although admittedly a greater knowledge and understanding of the genus being divided can make an important contribution to profit maximisation.

In industry, as in science, everything is classified: labour, liabilities, assets, annual costs, annual revenues, stores, output, even waste products, all are divided into species and classes by reference to relevant characteristics. The more scientifically compiled are such classifications the more surely they serve their intended purpose. By far the most startling results have been achieved in the materials stores. Scientific classification has revealed in some companies' stores that a few items had as many as ten or more different names, and that a stock had been kept under each, thus adding needlessly to the number of bins, the work of cataloguing, and most important, the total quantity of materials in store. The introduction of a scientific classification has sometimes resulted

[1] F. Bowen, *A Treatise on Logic*, Harvard, 1866, quoted by W. S. Jevons, *The Principles of Science*, London, 1874.

in annual savings in interest on working capital running into five figures.

Principles of classification

There are five main attributes of a good classification, viz:

1. The species and classes of the genus under review are mutually exclusive;
2. The division into species and classes is relevant to the purpose of the classification;
3. The classification is exhaustive;
4. The classification is flexible;
5. In a written classification, the wording is concise, complete and free of ambiguity.

It could be argued that these attributes are self-evident and a matter of common sense; yet the classifications that fail to possess one or more of them are legion.

Mutual exclusiveness

Many existing classifications have evolved over a period of many years, with the result that they bring chaos rather than order to the scene. Nobody has ever made a serious attempt to amend them or to compile new and sound classifications to take their place. One of the most common and most serious failings is a lack of mutual exclusiveness. *Ambiguous* is the word for this failing: the classification itself is ambiguous, and no matter how immaculate the wording used in the headings may be, it can never make good such a failing.

Consider a division of the genus manufactured goods. Such a division could very easily include the following headings:

Rubber goods
Gloves

on the grounds that each provides for the products of a distinct industry. But what is the result? Rubber gloves belong to either of the two classes and it is not clear which, so that the two classes are not mutually exclusive. The ambiguity arises from a failure to carry out the division by reference to a single characteristic. Rubber goods are divided from other products by reference to material; whereas gloves are divided by reference to function or shape.

One rule for achieving mutual exclusiveness is therefore to divide by reference to a single characteristic. It may be by material, or colour or smell or weight or any one of a number of qualities; but by only one at a time. If we decide to use material as the first division characteristic, then we must consider the problem of goods made of two or more materials, such as cars, aeroplanes, fur-lined rubber gloves, oil-heaters of pressed steel with brass fittings and leather boots with rubber soles. From this point of view, material as a first division characteristic is impracticable. If it were not, and it were adopted, then manufactured articles like gloves and boots would be scattered about the classification: under rubber, leather, cotton, asbestos, plastics, wool, silk and the rest as second, or third or *n*th division headings.

Would such a classification be useful? Would it not be better to have gloves and boots, and other kinds of finished goods each shown under a first division heading, with material shown as a lower division heading? Something like this:

Gloves:

Men's:	Women's:
Leather	Leather
Plastics	. . .
Rubber	

A classification like this with immediate function as the first division characteristic for the genus finished goods, could form part of a wider ranging classification showing a division for the genus raw materials and semi-finished goods with material as the first division heading.

Once a first division characteristic has been accepted, then different characteristics can be introduced for further division into sub-classes, provided that each class is divided by reference to a single characteristic. For instance, if the class is road motor vehicles, we might divide cars and motor-cycles by reference to the cubic capacity of the engine; buses and coaches by reference to seating capacity; and goods vehicles by reference to unladen weight.

Relevance

The second attribute of a sound classification is that it must be relevant. Relevance to the object of the classification is all important. Often, the object is the general one of distinguishing the species of a genus, the classes of each species, and so on down to the last physical detail.

There are many classes of orchid, and each individual orchid has physical differences from other individual orchids of the same class, such differences being due to chance and environment. Variations of this kind play no part in classification, which cannot by definition concern itself with individuals. Chance variations, even though they may be hereditary, are irrelevant. What is relevant are the characteristics common to the class, which may appear to be a question-begging kind of definition, but which in fact, merely serves to emphasise the difficulty of deciding where to draw the line. A class of moth varies in colour owing to differences in environment. Should colour be regarded as a relevant characteristic and the class be divided into two or more classes—not sub-classes, but classes each in its own right—by reference to colour? Since the class of moth is divided from other classes of moth by reference to a single characteristic or series of characteristics each taken in turn, there is no reason why colour should not be employed for the purpose, provided the various colours were distinct, but the several resulting classes would in fact be sub-classes of the class, though they may be regarded as existing on the same level as other classes of moth. If there is any question at all about the employment of colour, it is: would it serve any useful purpose? And this brings us back full circle. Is colour a relevant characteristic? Questions like these must have exercised the mind of Linnaeus many times when he was devising his classifications.

With classifications designed for internal use, there should be less difficulty in deciding what characteristics are relevant. For the stores, every physical difference would need to be taken into account in order to distinguish one type of material from another. Some types of manufactured materials, such as nuts, bolts and screws, may call for as many as five or more characteristics, taken one at a time to give a complete distinguishing specification, e.g.:

SCREWS:

- Wood:
  - Countersunk:
    - Steel:
      - No. 6 gauge:
        - ½ in.
        - ¾ in.
        - ...

Exhaustiveness

The attribute of exhaustiveness requires that a place must be provided for everything within the genus. That does not mean that a place should be specifically provided for every possible thing that could be regarded as belonging to the genus. In a factory stores, for instance, only the items which are kept in the store and which are likely to be kept, need to be provided for specifically. For a factory engaged in the production of car engines, the genus would be *parts of car engines*. A class for springs, divided into sub-classes, would be provided. But there would be no point in having a sub-class for hair springs for watches, or for any other kind of spring not used in car engines of the type, or in the machinery and equipment employed in the factory.

A device commonly used in general classifications to make them exhaustive, especially those devised for statistical purposes, is the residual heading. Thus we may have something like this:

WASHERS:

Iron
Steel
. . .
Other metals and alloys
Other materials

Here there are two residual headings, *Other metals and alloys*, and *Other materials*, which make the classification of washers exhaustive without going to great length. It is a useful device, no doubt, but it should be avoided in such classifications as those of materials in factory stores, where exact specification is of the first importance.

Flexibility

Flexibility, the fourth attribute of a sound classification, is closely related to exhaustiveness. It makes general provision for future changes and additions whether known or anticipated or not. Exhaustiveness makes full provision for the present, but where it anticipates, it can be said to embrace a measure of flexibility. However, anticipation makes specific provision; flexibility, general provision.

A rigid classification needs major amendment or may necessitate complete revision when changes in or additions to the genus take place.

A flexible classification may need only minor amendments and straightforward additions. Flexibility anticipates the kind of changes and additions, rather than the specific changes and additions, that will take place.

Flexibility is particularly important where a written classification is translated into a physical one, such as a factory stores, or where it forms the basis of a numerical code. Without flexibility in the physical classification, two or three additional classes may necessitate a complete rearrangement of the factory stores, and without flexibility in the numerical code, the whole code may have to be scrapped and a new one made.

Wording

Clarity and brevity provide the fifth and final attribute of a good written classification. A classification is not a glossary of definitions. The classifier has a right to suppose that the reader knows what a washer or a bolt or a circlip is, the meaning of self-tapping screw, lock-nut and valve-spring, and the difference between an electronic valve and a pressure-reduction valve.

All the classifier can be expected to do is to specify the characteristics that distinguish one class or sub-class from another. For this purpose, a single word is often all that is necessary, *washers* or *bolts* or *screws*, each of which conveys the distinguishing characteristics necessary for the classification.

Residual headings, which statisticians often refer to as rag bags, are a common stumbling block. They should not be used in a classification of factory stores; nor, where they are permissible, should they be used for a class that can easily be precisely specified. A temptation to use the word *Other* is always present. The reader, for his part, is inclined to ask 'other what?' and 'other than what?' Insetting designed to give the answers to such questions needs to be employed with care. In some official classifications, it is necessary to use a ruler and set-square to find the answers. It is a simple matter to answer the former question by inserting the appropriate noun after *Other: 'Other washers'*, *'Other bolts'* and so on. The latter question is not so easy to answer; but the answer to the former goes a long way towards it.

Phrases denoting exceptions in headings are best inserted in brackets to avoid ambiguity. A heading like 'Soap, excluding shaving soap, soap powders and soap flakes' is ambiguous since it is not clear whether 'soap powders and soap flakes' are qualified by 'excluding' or not.

Order of Headings

In all classifications, class order is fundamental to their purpose. Should the classification of sheet-metal products in a factory stores show material as the first division, or the type of product? For instance, the classes could be shown:

Steel
  Expanded metal
    $\frac{1}{4}$ in mesh
    $\frac{1}{2}$ in mesh
    ...
Perforated metal
    ...
Aluminium
  Expanded metal
    ...

or, alternatively:

Expanded metal
  Steel
    $\frac{1}{4}$ in mesh
    ...
  Aluminium
    $\frac{1}{4}$ in mesh
    ...
Perforated metal
  Steel
  ...

Neither of these is absurd; and the choice must depend upon the purpose of the classification. What would be absurd for most purposes would be a prior division made by reference to size, such as:

METAL PRODUCTS:
  $\frac{1}{2}$ in mesh
    Expanded metal
      of steel
      ...
    Wire netting
      ...
  $\frac{3}{4}$ in mesh
    Expanded metal
      ...

Classification by reference to size, at whatever level, has a natural order. Not so, classifications by reference to most other characteristics. It may be the natural order of things to begin a commodity classification, such as the *Export List*, with live animals, proceed to animal products—carcass meat, milk, eggs—thence to agricultural and forestry products, on to other raw materials—minerals and metals—and terminating with manufactured goods.

Beyond that, the classifier runs into difficulties if he seeks a natural order. Have eggs, milk and meat a natural order? Have pork, mutton and beef a natural order? And fish—one could and perhaps should classify all salt-water fish together and fresh-water fish together, with eels and salmon in a separate category. But there does not seem to be any natural reason why salt-water fish should appear before fresh-water fish, or fresh-water fish before salt-water fish.

Heading order has been a subject of some controversy among classifiers for a long time. Some argue that there is always a natural order of all things; others that there is not—that there may be a natural order in some things, but by no means all. Arbitrary as it may seem, alphabetical order is preferable in divisions where logic has to be stretched to produce what is claimed to be a natural order. Alphabetical order is used extensively in many official classifications such as the *Export List*, and to good purpose: it is a great convenience to their many users.

Classification is a complex subject. It is one that the management statistician cannot afford to neglect: and indeed, he may derive a certain amount of satisfaction from solving the many problems that he is certain to encounter. In a classification of engineering products, should a cobbler's sewing machine be classified with sewing machines or with boot-and-shoe-making machinery? How will the classifier make his choice clear in the wording of the classification? Should a bicycle chain be classified with transmission chain or with cycle parts? On what grounds will he defend his choice? And how will he make that choice clear in his written classification? I would say from my own experience in this field that making a choice and justifying it are simple matters; the real problem is to make the choice clear in the classification. It is worth mentioning here, that end-use is no criterion. If the cobbler's sewing machine were the same as other sewing machines, it would properly be classified to sewing machines. But since it has some fundamental physical differences, a choice exists, and a decision must be made.

Perhaps the problem arises from the terminology. If the cobbler's sewing machine were called a leather-stitching machine, there would be no problem. Even so, whenever a problem like this arises, the classification does not possess the first attribute of a good one, that of mutual exclusiveness. The question is then: is the classification divided by reference to a single characteristic ? and the answer must be that it is not. Sewing machines are classed by reference to immediate function, whereas boot and shoe machinery is classed by reference to using industry. All types of machinery may be described by reference to immediate function, but some, such as looms and potters' wheels, are specific to definite using industries and may be so classified. It is the types of machinery that have a description which implies a general use that create the difficulty. Either categories described by reference to immediate function must be given priority over categories described by using industry, or *vice versa.* There are two obvious ways of achieving this: one by giving specific instructions to users, and the other by adapting the tree of Porphery method, i.e.:

Machinery defined by immediate function:

Looms
Potters' wheels
Sewing machines
. . .

Other:

Machinery defined by using industry:

Boot and shoe machinery
Pottery making machinery
Textile machinery
. . .

Perhaps the ultimate solution is to abandon the using-industry classes and to classify all machinery by reference to immediate function. However, such a solution would be far from satisfactory for most purposes: it would involve many more headings, many of minor importance and for somewhat remote processes; and it might mean the loss of some useful statistical totals relating to machinery for particular using industries, albeit those totals could rarely if ever relate to the whole of the machinery destined to be used in the industries specified.

REPORT WRITING

Reports are one of the more important methods of internal communication. They form part of the management information service, and as such they should meet the requirements of the recipients within the report writer's terms of reference.

The main attributes of a good report may be stated as follows:

1. It is readable;
2. It is interesting;
3. It presents the facts and conclusions clearly and concisely;
4. The facts are stated objectively. The conclusions are logical inferences drawn from the facts as stated; and where the report writer expresses an opinion, he makes it clear that it is an opinion and states whose opinion it is;
5. In general, jargon is avoided, and technical terms are used only where it is known that the reader will understand them; and
6. The opening paragraphs state the object of the underlying investigation; the middle paragraphs contain and discuss the ascertained and verified facts, and the concluding paragraphs give the conclusions and recommendations.

Making a report readable, interesting, clear and concise comes of a skill that only long practice and experience can impart. Like classifying, report writing calls for a marshalling of the facts, and some considerable measure of concentrated thought. First drafts may be dictated by those with the facility; but in general, it is a pen-and-paper exercise. *Ad hoc* reports usually call for more thought and concentration than reports belonging to a series under a single general heading. With the latter, the format or layout has usually become a set pattern, from which necessary departures may be rare; and the kinds of facts being sought are known.

There are many pitfalls for the unwary, of which a few, but by no means all, are considered below.

Pitfalls

First there are the pitfalls which may arise from either wrong thinking or careless writing. Most of the reasoning required for report writing is deductive. A great deal of deduction takes the form of the syllogism, usually with one of the premises omitted. The classic complete syllogism is:

All metals are elements
Iron is a metal
Therefore iron is an element

which reduces to 'Iron is a metal and therefore an element'. The implication of the reduced form is the major premise, so called, of the complete syllogism, i.e. that all metals are elements. It is a good idea when arguing on lines similar to this to test the implication of the argument—in effect to determine the major premise underlying it, and ask, 'Is it valid?' In the classic example, one might question the validity of the major premise: is it true that all metals are elements? Clearly the argument refers only to elemental metals and does not cover alloys such as brass and pewter. Examples of incomplete syllogisms or arguments of a syllogistic type are frequent. 'This china comes from Staffordshire and is therefore of good quality' implies that all china that comes from Staffordshire is of good quality, which is the implicit major premise.

The non sequitur

A syllogistic argument that fails to satisfy the major-premise test is called a *non sequitur*, which means that the conclusion does not follow from the stated premise or premises. If it is not true that china from Staffordshire is all of good quality, then it does not follow that because a certain piece comes from Staffordshire it is of good quality.

There are other forms of the *non sequitur*, such as the careless use of such conjunctive adverbs as *therefore*, *hence* and *consequently*. A common one resulting from careless thinking is the conclusion that is too precise for the premises stated, e.g.:

> The Board's original off-form cocoa price for the 1949–50 season was set at a period when the world price was low, and was therefore between 15 and 20 per cent lower than that originally offered for the 1948–49 season.

The non causa

The *non causa* or false cause is similar to the *non sequitur* and is often confused with it, largely because the same conjunctive adverbs are used. Arguing from cause to effect is induction whereas arguing from premise to inference is deduction. Major premises, whether stated or understood, are sometimes determined by induction. That all china from Staffordshire is of good quality could be proved only by induction. Other major premises are true by definition, that all metals are elements being an example; but since the term *metals* clearly means elemental metals, the premise is a tautology. The false cause is a lapse of logic which is all too common, and it is not always easy to determine whether any particular case is due to careless thought or careless writing.

> Four-power control would be necessary to see that too much currency did not circulate, thus causing inflation.

The person who wrote this was as guilty of woolly thinking as of careless writing.

Misuses of the Negative

There is little fear that a fairly well educated person could ever use a double negative in writing even though he may sometimes say such things as 'I couldn't hardly do it.' There is nothing ambiguous about this kind of grammatical lapse; but the legitimate use of the double negative in such phrases as 'not infrequently', 'not impossible' and 'not unlikely' can give rise to doubt in the mind of the reader. Does one negative cancel out the other? Does 'not infrequently' mean 'frequently'? In cases like this the writer is being deliberately vague, largely because he is not sure of his facts. It is a type of construction best avoided in business reports.

Grammatical double negatives go beyond this kind of phrase:

> This point is important, since not a few writers not infrequently look for a fixed rather than a flowing order of discourse.

This contains two double negatives—a double double negative, so to speak: and as a result, the reader is liable to find the sentence perplexing '... some writers often look ...' would probably convey the writer's meaning exactly without confusing the reader.

A few more examples, with comments, of the misuse of the negative are given below. It should be mentioned that these examples, like those above, are extracted from the works of professional writers, academics and publicists.

> This remark was not concerned specifically with grinding operations, but also covered separating and screening.

Slovenly is not too strong an epithet for this passage. The negative appears to be quite superfluous: 'This remark was concerned with separating and screening as well as with grinding.'

> It will not be, as Mr Jay at one time appeared to be claiming, a tax on luxury spending.

What did Mr Jay claim, that it would be a tax or not be a tax on luxury spending? This is a common kind of ambiguity, of which another example is:

> The four foreign ministers did not meet this afternoon as expected.

A similar kind of ambiguity is the negative followed by 'in order to' or 'in order that'.

> It should perhaps be explained that machines did not create man in order that there may be operators to control them.

And another is the negative followed by because:

> People are not dumb because they lack mental equipment; they are dumb because they lack an adequate method.

A logical interpretation of this is: 'It is because people lack an adequate method that they are dumb; it is because they lack mental equipment that they are not.' A sensible interpretation is: 'People are dumb, not because they lack mental equipment, but because they lack an adequate method.' (Dumb in this context is American slang for *stupid* or *incompetent.*)

A more obscure example of '... not ... because ...' is:

> Jugoslavia will not renounce these principles, in any circumstances, because of pressure from outside.

Whether Jugoslavia will renounce or not renounce these principles because or not because of pressure from outside is not clear.

The Queen's English

Each profession has its own peculiar style of English, by far the most readable, clear and concise being that of the legal profession. It tends to be formal in style and to avoid modern idiom. Whether the readability and clarity come of the lawyer's work of interpreting the most difficult of English, the language of the Statutes, is a matter of conjecture. There would be no point in an interpretation that could not be understood. Lawyers' English of the kind is found in the write-ups of court cases in *The Times, The Certified Accountants Journal, The Accountant* and similar periodicals, but perhaps more particularly in the summing up made by judges. An example of the latter, which happens to be relevant to our subject, is as follows:

> What does the Act mean? What is the true construction? It is impossible to escape deciding that question of law by saying the Act is so slovenly and so unintelligible that it is impossible satisfactorily to ascertain and declare its meaning. If it were competent to a court of law to censure the Legislature, or if any useful purpose could be served by censuring the Legislature of 1842 to 1853, no censure could be too strong, I think, for having expressed an Act, and a Taxing Act, in language so involved. But there it is. A court cannot say it means nothing, and cannot be construed at all. The Court must, as best it can, arrive at some meaning of the language which bears upon the particular case before it for decision.[1]

Scarcely anything could be clearer, more concise and more readable than that.

Apart from the language of the Statutes, the other extreme is found in the language of the computer experts, whose ability to express a simple thought in plain English appears to be almost completely lacking. Consider the following extracts from an article published in a USA magazine devoted to business finance.

1. The Macro-level analysis draws upon management's experience with the operating systems and the system's group expertise in computer operations, data processing, anad analysis to translate information needs into defined requirements which can be interpreted in terms of specific data processing activities.
2. Major decision areas and decision branches must be identified and related to the conceptual framework.
3. The framework must facilitate identification of specific criteria or models for decision appropriate to the structure and objectives of the system.

This kind of jargonised English may be all very well in an article, paper or report addressed to other experts in electronic data processing; but in an article or report addressed to accountants and managers, it is indefensible. Fortunately, statisticians are not given to such writing;

[1] Lord Wrenbury, quoted from 'Tax Terminology—*xxv*', *The Certified Accountants Journal*, July–August 1971.

but it could easily happen. Oddly enough, there appears in the same issue of the periodical an article by a statistician on EDP written in clear English.

Throughout history, men have been concerned with communication, and the use of clear language. The ancient Greeks concentrated on logic and clear argument; but there have been more forthright aspirations to use understandable language. In his First Epistle to the Corinthians, the Apostle Paul wrote:

> For if the trumpet give an uncertain sound, who shall prepare himself to the battle?
> So likewise ye, except ye utter by the tongue words easy to be understood, how shall it be known what is spoken? for ye shall speak into the air.
> Therefore if I know not the meaning of the voice, I shall be unto him that speaketh a barbarian, and he that speaketh shall be a barbarian unto me.
>
> I Cor xiv: 8, 9, 11.

Half the mysteries of EDP may lie in the esoteric language used by the experts. Perhaps some computer failures are attributable to it, too.

OTHER MATTERS IN BRIEF

Much of the management statistician's work varies according to the kind of business. The management statistics appropriate to a bank are entirely different in many respects from those of a building contractor, and the management statistics appropriate to a manufacturer are entirely different from those of an insurance company or life assurance office. Consider, for instance, the annual budget. A manufacturer or distributor may summarise his budget like this:

	£	£
Sales proceeds		
Annual costs		
Direct		
Overheads		
Profit before tax		

It is questionable whether either a bank or an insurance company could put a figure to sales proceeds. They both sell services, the gross revenue of one consists largely of interest on loans and investments; and of the other, largely of premiums. It is also questionable whether the annual costs of either could be pigeon-holed so neatly. Insurance statistics are highly specialised, and the companies employ specialist statisticians, known as actuaries. Actuaries have two professional examining bodies in the United Kingdom, viz: the Institute of Actuaries, Staple Inn Hall, High Holborn, London E.C.1; and the Faculty of Actuaries in Scotland, 23, St Andrew Square, Edinburgh. Both grant qualifications which are generally recognised by insurance companies.

Statistics relating to the introduction of a new brand of product or a new service also vary according to the kind of business carried on. They are closely related to and indeed ultimately form part of the statistics for budgeting.

In these days of liquidity problems, an important task for the statistician is to forecast the liquidity position of the company week by week or month by month during the next year, to bring the forecasts up to date as frequently as possible, and to compare his forecasts with the actual cash position as time passes. His weekly or monthly report to top management on this subject could usefully bring to light weaknesses in such spheres as cost control and the bills receivable office. For this purpose, figures of the inflow and outflow of cash during the period and accruing debits and credits, as well as the position at the end of the period would be useful. Some ratios, too, might be helpful in the same direction, e.g., the ratio of cash inflow to accruing debtors in the period.

There are other spheres of business activity where the management statistician can be of assistance, or where he can obtain statistical information of use to top management. Indeed it is difficult to imagine any sphere where the management statistician could not play a part. And the more he knows about the business, its products or services and its financial, statistical, costing and other systems, the more useful is the part he can play.

Index